T0300505

Deborah L. VanOverbeke, PhD
Editor

Handbook of Beef Safety and Quality

Pre-publication REVIEWS, COMMENTARIES, EVALUATIONS . . .

"In its totality, the *Handbook of Beef* Safety *and Quality* is a well-crafted blend of comprehensive reviews of literature by 'top of their game' scientists. A 'must-read' for anyone interested in updating his or her knowledge and understanding of the industry's efforts to improve the safety and quality of beef, and of the scientific community's efforts to understand the mind-set of those who purchase, merchandize, and/or consume beef products."

Gary C. Smith, PhD
Professor, Department of Animal Sciences, Colorado State University

"This is a very useful and valuable resource for utilization in the classroom. Currently there exists no comprehensive overview for these all-important emerging issues within the beef industry. The text provides practical and informative insight. Students who want to work in the beef industry at any level need to have a thorough understanding of the various aspects related to beef safety. The authors provide that in a cogent and readable manner.

Quality considerations are increasingly important in this business—a shift in what has long been a traditional commodity business. As such, students will find the beef quality section especially constructive. Many animal science students receive limited training from a business perspective and fall short of the needed curriculum with respect to quality management. This text provides a useful framework from a historical perspective about where the beef industry has been and fills in the educational gap regarding quality management not available in most animal science curriculums."

Nevil C. Speer, PhD, MBA
Professor, Animal Science, Department of Agriculture, Western Kentucky University

More pre-publication
REVIEWS, COMMENTARIES, EVALUATIONS . . .

"*Handbook of Beef Safety and Quality* is a comprehensive review of factors that affect beef safety and quality throughout the production chain. The book details significant factors that can positively or negatively impact safety and quality. Instructors, students, and industry personnel should be able to utilize the book as a stand-alone reference for beef safety and quality or as a supplemental text or reference to traditional meat science textbooks. The beef safety section of the book compiles all information and facts needed to clearly understand the factors influencing beef safety and how the industry is working to continually improve product safety from feedlot to fork. The book is a first of its kind because of its comprehensive nature to address the major factors impacting beef quality and safety."

Gretchen Hilton, PhD, MS
Assistant Professor, Meat Science,
Oklahoma State University

Haworth Food & Agricultural Products Press™
An Imprint of The Haworth Press, Inc.
New York • London • Oxford

NOTES FOR PROFESSIONAL LIBRARIANS AND LIBRARY USERS

This is an original book title published by Haworth Food & Agricultural Products Press™, an imprint of The Haworth Press, Inc. Unless otherwise noted in specific chapters with attribution, materials in this book have not been previously published elsewhere in any format or language.

CONSERVATION AND PRESERVATION NOTES

All books published by The Haworth Press, Inc., and its imprints are printed on certified pH neutral, acid-free book grade paper. This paper meets the minimum requirements of American National Standard for Information Sciences-Permanence of Paper for Printed Material, ANSI Z39.48-1984.

DIGITAL OBJECT IDENTIFIER (DOI) LINKING

The Haworth Press is participating in reference linking for elements of our original books. (For more information on reference linking initiatives, please consult the CrossRef Web site at www.crossref.org.) When citing an element of this book such as a chapter, include the element's Digital Object Identifier (DOI) as the last item of the reference. A Digital Object Identifier is a persistent, authoritative, and unique identifier that a publisher assigns to each element of a book. Because of its persistence, DOIs will enable The Haworth Press and other publishers to link to the element referenced, and the link will not break over time. This will be a great resource in scholarly research.

Handbook of Beef Safety and Quality

CRC Press is an imprint of the
Taylor & Francis Group, an informa business

HAWORTH FOOD & AGRICULTURAL PRODUCTS PRESS™

Mineral Nutrition of Crops: Fundamental Mechanisms and Implications by Zdenko Rengel

Conservation Tillage in U.S. Agriculture: Environmental, Economic, and Policy Issues by Noel D. Uri

Cotton Fibers: Developmental Biology, Quality Improvement, and Textile Processing edited by Amarjit S. Basra

Heterosis and Hybrid Seed Production in Agronomic Crops edited by Amarjit S. Basra

Intensive Cropping: Efficient Use of Water, Nutrients, and Tillage by S. S. Prihar, P. R. Gajri, D. K. Benbi, and V. K. Arora

Physiological Bases for Maize Improvement edited by María E. Otegui and Gustavo A. Slafer

Plant Growth Regulators in Agriculture and Horticulture: Their Role and Commercial Uses edited by Amarjit S. Basra

Crop Responses and Adaptations to Temperature Stress edited by Amarjit S. Basra

Plant Viruses As Molecular Pathogens by Jawaid A. Khan and Jeanne Dijkstra

In Vitro Plant Breeding by Acram Taji, Prakash P. Kumar, and Prakash Lakshmanan

Crop Improvement: Challenges in the Twenty-First Century edited by Manjit S. Kang

Barley Science: Recent Advances from Molecular Biology to Agronomy of Yield and Quality edited by Gustavo A. Slafer, José Luis Molina-Cano, Roxana Savin, José Luis Araus, and Ignacio Romagosa

Tillage for Sustainable Cropping by P. R. Gajri, V. K. Arora, and S. S. Prihar

Bacterial Disease Resistance in Plants: Molecular Biology and Biotechnological Applications by P. Vidhyasekaran

Handbook of Formulas and Software for Plant Geneticists and Breeders edited by Manjit S. Kang

Postharvest Oxidative Stress in Horticultural Crops edited by D. M. Hodges

Encyclopedic Dictionary of Plant Breeding and Related Subjects by Rolf H. G. Schlegel

Handbook of Processes and Modeling in the Soil-Plant System edited by D. K. Benbi and R. Nieder

The Lowland Maya Area: Three Millennia at the Human-Wildland Interface edited by A. Gómez-Pompa, M. F. Allen, S. Fedick, and J. J. Jiménez-Osornio

Biodiversity and Pest Management in Agroecosystems, Second Edition by Miguel A. Altieri and Clara I. Nicholls

Plant-Derived Antimycotics: Current Trends and Future Prospects edited by Mahendra Rai and Donatella Mares

Concise Encyclopedia of Temperate Tree Fruit edited by Tara Auxt Baugher and Suman Singha

Landscape Agroecology by Paul A. Wojtkowski

Concise Encyclopedia of Plant Pathology by P. Vidhyasekaran

Molecular Genetics and Breeding of Forest Trees edited by Sandeep Kumar and Matthias Fladung

Testing of Genetically Modified Organisms in Foods edited by Farid E. Ahmed

Handbook of Beef Safety and Quality edited by Deborah L. VanOverbeke

Handbook of Beef Safety and Quality

Deborah L. VanOverbeke, PhD
Editor

CRC Press is an imprint of the
Taylor & Francis Group, an informa business

Reprinted 2010 by CRC Press
CRC Press
6000 Broken Sound Parkway, NW
Suite 300, Boca Raton, FL 33487
270 Madison Avenue
New York, NY 10016
2 Park Square, Milton Park
Abingdon, Oxon OX14 4RN, UK

The Haworth Press, Inc., 10 Alice Street, Binghamton, NY 13904-1580.

PUBLISHER'S NOTE
The development, preparation, and publication of this work has been undertaken with great care. However, the Publisher, employees, editors, and agents of The Haworth Press are not responsible for any errors contained herein or for consequences that may ensue from use of materials or information contained in this work. The Haworth Press is committed to the dissemination of ideas and information according to the highest standards of intellectual freedom and the free exchange of ideas. Statements made and opinions expressed in this publication do not necessarily reflect the views of the Publisher, Directors, management, or staff of The Haworth Press, Inc., or an endorsement by them.

Cover design by Lora Wiggins.

Library of Congress Cataloging-in-Publication Data

Handbook of beef safety and quality / Deborah L. VanOverbeke, editor.
p. cm.
ISBN-13: 978-1-56022-323-8 (hard 13 : alk. paper)
ISBN-10: 1-56022-323-5 (hard 10 : alk. paper)
ISBN-13: 978-1-56022-324-5 (soft 13 : alk. paper)
ISBN-10: 1-56022-324-3 (soft 10 : alk. paper)
1. Beef industry—United States. 2. Beef industry—United States—Safety measures. 3. Beef—Quality—United States. 4. Beef—Quality—Health aspects—United States. 5. Beef cattle—United States. I. VanOverbeke, Deborah L.

HD9433.U4H36 2007
363.16'29—dc22

2006031195

CONTENTS

About the Editor ix

Contributors xi

Preface and Acknowledgments xvii

PART I: BEEF SAFETY

Chapter 1. Introduction: The Safety of Beef 3

John Paterson

Consumer Demands for Beef Safety 3
Common Foodborne Diseases 6
Why Foodborne Pathogens Are of Concern to the Beef Industry 11
Summary 15

Chapter 2. Preharvest Beef Safety: Production Management and Pathogen Control 19

Guy H. Loneragan
Mindy M. Brashears

Preharvest Control of Pathogens: Is it Possible and Should it Even Be Considered? 19
Pathogens of Interest: Past and Future 22
Direct-Fed Microbials 27
Immunomodulation 30
Summary 33

Chapter 3. Beef Safety During Slaughter, Fabrication, and Further Processing **39**

Sally L. Flowers Yoder
Margaret D. Hardin
William R. Henning
Catherine N. Cutter

Introduction 39
Chemical and Physical Hazards 43
Biological Hazards 47
Antimicrobial Interventions 54
Further Processing 67
Summary 72

PART II: BEEF QUALITY

Chapter 4. The Quality Revolution **85**

Thomas G. Field
Deborah L. VanOverbeke

Introduction 86
Overview of Total Quality Management 89
The Role of Professional Standards in International Trade 91
Defining the Role of Quality in Enhancing Beef Demand 94
Moving the Industry from Commodity to Value-Added 97
Developing Quality-Focused Supply Chains 98
Summary 99

Chapter 5. Preharvest Beef Quality **101**

Robert A. Smith

History of Preharvest BQA 101
Residue Avoidance 103
Injection-Site Lesions 109
Care and Handling of Beef Cattle 113
Relationship of Disease to Performance and Beef Quality 120
Other BQA Issues 122
The Future of Preharvest BQA 122

Chapter 6. Beef Carcass Quality **125**

Jeff W. Savell
Carrie L. Adams Mason
F. Danielle Espitia
Diana Huerta-Montauti
Kristin L. Voges

Introduction 125
Relationship Between Carcass Quality and Consumer Acceptability 137
Relationship Between Mouthing and Skeletal Maturity 140
Summary 143

Chapter 7. Sensory Attributes and Quality **147**

Rhonda Miller

Introduction 147
Methods Used in Beef Sensory Evaluation 148
Beef Sensory Attributes 155
Factors Impacting Beef Sensory Attributes and Quality 160
Palatability of Different Beef Muscle 176
Summary and Conclusions 179

Chapter 8. Beef Quality, Beef Demand, and Consumer Preferences **187**

Wendy J. Umberger

Introduction 187
Demand and Consumer Beef Demand Determinants 188
Consumer Beef-Demand Models 190
What Is Beef Quality and How Do Consumers Develop Their Perceptions of Quality? 196
Consumer Preferences for Quality Attributes and Cues 203
Summary 208

Glossary **215**

Index **223**

ABOUT THE EDITOR

Deborah L. VanOverbeke, PhD, received her BS in animal science from the University of Nebraska-Lincoln and her MS and PhD in meat science from Colorado State University. Dr. VanOverbeke is currently on the faculty at Oklahoma State University, where her focus for teaching and research emphasizes on pre- and postharvest beef-quality issues, traceability, and quality-assurance programs. Prior experience included serving on the faculty at the University of Minnesota and on the staff of the Nebraska Cattlemen's Association.

Handbook of Beef Safety and Quality

doi:10.1300/5640_a

CONTRIBUTORS

Dr. Mindy M. Brashears graduated from Texas Tech University with her BS in 1992 (food technology), MS from Oklahoma State University in 1994 (food science), and PhD (food science) from Oklahoma State University in 1997. After completing her PhD she was an assistant professor at the University of Nebraska in the Department of Food Science and Technology from 1997 to 2001. She is currently an associate professor and director of the International Center for Food Industry Excellence at Texas Tech University. Her research focuses on interventions in pre- and postharvest environments and on the emergence of antimicrobial drug resistance. She teaches courses in food microbiology and food safety and offers industry-training opportunities in Hazard Analysis and Critical Control Points (HACCP), advanced HACCP, food sanitation, recalls, and food security.

Dr. Catherine N. Cutter received her BS and MS from the University of Connecticut in pathobiology and a PhD in food technology from Clemson University. She was employed as a research microbiologist with the USDA-Agricultural Research Service, U.S. Meat Animal Research Center in Clay Center, Nebraska, for over seven years. In 1999, she joined the faculty of the Department of Food Science at Penn State, where she is an associate professor and food safety extension specialist who specializes in muscle foods. Dr. Cutter's current research interests include examining methods or interventions to reduce or eliminate pathogenic and spoilage bacteria on muscle foods (meat, poultry, seafood) during slaughtering, processing, and fabrication, including the indirect application of antimicrobials to muscle foods through packaging. Her extension/teaching programs include food safety education for food processors,

Handbook of Beef Safety and Quality

doi:10.1300/5640_b

sanitation, general food microbiology, irradiation, control of *Listeria monocytogenes,* and HACCP for meat and poultry processors.

F. Danielle Espitia is a former teaching/research assistant in the Department of Animal Science at Texas A&M University and received her MS in animal science in 2006. She currently works as the quality-assurance manager for DOS Projects in Saginaw, Texas.

Dr. Thomas G. Field received all three of his degrees from Colorado State University. A native Coloradoan, he teaches introduction to food animal science, beef production and industry, beef feedlot management systems, and family ranching. He is also coordinator of the seedstock merchandising team, manages performance data for the seedstock cattle in the RI herd, and serves on the management team for the livestock program at ARDEC. His work focuses in beef cattle management systems with an emphasis on improving the profitability of ranching. A frequent speaker at state, national, and international beef cattle educational events, Dr. Field is also the co-author of *Scientific Farm Animal Production* and *Beef Cattle Management Decisions.*

Dr. Margaret D. Hardin is the director of food safety and quality assurance with Boar's Head Provisions Co., Inc. Prior to her employment at Boar's Head, she held positions as the director of food safety at Smithfield Packing/Gwaltney Foods, Sara Lee Foods, and the National Pork Producers Council and as a research scientist and HACCP instructor with the National Food Processor Association in Washington, DC. Her efforts have been directed in areas of food safety, research, HACCP, and sanitation to protect the public health and assure the microbiological quality of food. Margaret has been a member of both the National Advisory Committee on Microbiological Criteria for Food and the National Advisory Committee on Meat and Poultry Inspection. She is also a member of several professional and scientific groups, including the International Association for Food Protection, the Institute of Food Technologists, American Meat Science Association, and the American Society for Microbiology.

Dr. William R. Henning obtained his MS in animal science and PhD in meat science from the University of Kentucky. After two years with Swift and Co., Dr. Henning was an assistant professor at Western Illinois University and then joined the faculty at the Pennsylvania State University as an extension meats specialist in 1982. He has conducted numerous meat-processing workshops and carcass-evaluation programs for the livestock industry. He has been involved with educating beef and dairy producers about meat safety and currently serves on the Advisory Board for the NCBA Beef Quality Assurance program, as well as the NCBA Animal Health Network. His extension and outreach responsibilities are primarily in the areas of food safety, meat processing, and meat quality. Dr. Henning has conducted applied research in meat safety, including identifying sources of contamination, developing intervention strategies for preventing contamination with *E. coli* O157:H7 in beef, and identifying the potential presence of *M. paratuberculosis* in beef.

Diana Huerta-Montauti is a former teaching/research assistant in the Department of Animal Science at Texas A&M University and received her MS in animal science in 2006.

Dr. Guy H. Loneragan received his BVSc at the University of Sydney, Australia, in veterinary science and his MS and PhD at Colorado State University. His academic areas of interest include veterinary medicine and epidemiology with specific interest in feedlot animal health, morbidity and mortality modeling, and food safety. Current research includes preharvest epidemiology and ecology of *E. coli* O157, preharvest ecology of antimicrobial drug resistance, and epidemiology of BSE in North America.

Carrie L. Adams Mason is a former teaching/research assistant in the Department of Animal Science at Texas A&M University and received her MS in animal science in 2006. She currently works as the program manager for the American Meat Science Association in Savoy, Illinois.

Dr. Rhonda Miller is a professor in meat science in the Department of Animal Science at Texas A&M University. She received her BS, MS, and PhD from Colorado State University. Dr. Miller teaches graduate and undergraduate courses and directs the Sensory Testing facility. Her current research includes antemortem and postmortem factors affecting the composition, palatability, and quality of red meat, and the development of value-added red meat products.

Dr. John Paterson is the extension beef specialist at Montana State University, where he has primary responsibilities for educational programs in beef cattle management and nutrition, food safety, and animal identification and traceability. Dr. Paterson is a 1979 graduate of the University of Nebraska, and has held positions at the University of Missouri and Montana State University.

Dr. Jeff W. Savell is Regents Professor and E.M. "Manny" Rosenthal Chairholder in the Department of Animal Science at Texas A&M University. He teaches undergraduate and graduate courses, conducts research on quality, safety, and value, and serves the public through many meat science outreach programs.

Dr. Robert (Bob) A. Smith is a 1976 graduate of the College of Veterinary Medicine at Kansas State University, holds an MS from Kansas State, and is a Diplomate, American Board of Veterinary Practitioners. Dr. Smith retired in 2002 and presently is at Veterinary Research and Consulting Services, LLC, in private practice limited to stocker and feeder cattle. As a faculty member at Oklahoma State University's College of Veterinary Medicine, he served as the director of Veterinary Extension and coordinator of Beef Cattle Health Studies at the Willard Sparks Beef Cattle Research Center. He is a nationally recognized authority in stocker and feeder cattle production medicine and applied bovine respiratory disease research.

Dr. Wendy J. Umberger received her BS in animal science and MS in economics from South Dakota State University, and her PhD in agricultural economics from the University of Nebraska-Lincoln. Dr. Umberger is currently on the faculty at Colorado State University, where her focus for teaching and research emphasizes consumer willingness to pay for food quality attributes, traceability, agricultural production economics, marketing, and integrated resource management.

Kristin L. Voges is a teaching/research assistant in the Department of Animal Science at Texas A&M University and is pursuing a PhD in animal science.

Sally L. Flowers Yoder earned her MS in animal science and PhD in food science from the Pennsylvania State University. As a graduate assistant, most of her research was focused on the microbiological safety of red meats processed in very small establishments. She currently lives in Starkville, Mississippi, with her husband, Tim, and daughter, Hazel.

Preface and Acknowledgments

The concept for this text was put into place by Haworth Press. Beyond the concept, the actual direction for the text was developed with the help of the American Meat Science Association membership in response to a survey requesting their input for the topics important to address in a text titled *Handbook for Beef Safety and Quality.* With the input, the text was put together in two sections: Beef Safety and Beef Quality. In each section, chapters address preharvest controls for safety and quality, harvest controls for safety and quality, and postharvest controls for safety and quality. The hope is that this text can be used as a guidance tool in the beef industry as well as by students who are learning more about the impacts of production, harvest, and processing on beef safety and quality as the product exists in the marketplace today.

Appreciation is expressed to each of the authors of the chapters in this book. It is because of their dedication to the beef industry and their desire to help explain complex topics to a vast audience that this text exists. They have spent hours in the development of their chapters and their expertise is appreciated. Also, many thanks to Kris Novotny, who helped coordinate the glossary of terms from the chapters, and to Micca Sullivan for assisting with the final preparation of the manuscript. And, last, but not least, to my husband, Kirk, for his unending love and support during the process of preparing the finished product.

Handbook of Beef Safety and Quality

doi:10.1300/5640_c

PART I: BEEF SAFETY

Chapter 1

Introduction: The Safety of Beef

John Paterson

CONSUMER DEMANDS FOR BEEF SAFETY

The "produce-and-then-sell" mentality of the cattlemen who produce commodity beef is rapidly being replaced by the strategy of first asking consumers what they want as attributes in their beef products, and then creating or manufacturing it (Smith, 2003). Schroeder (2003) said, "The opportunity for industry-wide profitability will be absolutely dependent upon . . . a clear-eyed, relentless focus on the consumer, and what is required for her or him to persistently spend more money on beef" (pp. 4-5).

Value is created when the product that is generated meets or exceeds the expectations of the customer or consumer. This old saying is true today.

At the consumer level, quality attributes such as tenderness, flavor, and portion size are important. At the production level, concerns are focused on such things as performance, health, and predictability all through the system. Beef-quality factors are affected by management (Texas Beef Quality Assurance Manual, 2005, available at www.texasbeefquality.com/) while demand is affected by both the consumers' perception of various product attributes of beef and demographics of the consumers themselves. The product attributes that are known to affect beef demand include (1) food safety (is this product safe?), (2) nutrition (is it good for me?), (3) palatability (is it tender and tasty?), (4) convenience (is it convenient to prepare?), and (5) the

Handbook of Beef Safety and Quality

doi:10.1300/5640_01

relative price when compared to competing meats. These product attributes have frequently been called the demand drivers because they have been shown to drive the demand for beef (Genho, 2003).

Dr. Harlan Ritchie from Michigan State University said, "The beef industry has changed more in the past six years than it did in the previous sixty" (Oldham, 2002). Paterson (2002) concurred by writing, "It took 20 years to turn beef demand around because the industry's price system failed to provide signals—premiums for doing things right; discounts for doing things wrong—back to producers" (p. 25).

The beef industry has been said to be moving from "production-driven" to becoming "consumer-driven," and this is simply another way of saying "end-user responsive" (Smith, 2001a). Smith (2001b) concluded that producers must know the difference between a "customer" and a "consumer." Examples of such differences include the following:

- A feedlot operator can say that he or she is satisfying the customer by providing the most desirable carcass (in terms of quality grade, yield grade, carcass weight, rib-eye area, defect criteria) as defined by a packer grid, but may not be satisfying the consumer (the person who eats the beef).
- To be consumer driven means that cattle producers can no longer just provide what they think is best (or easiest or most economical) and expect the world to come begging for more.
- Rather, being consumer driven means that at each critical juncture in the beef production sequence, consideration must be given to what the consumer wants and is willing to buy.

Table 1.1 briefly summarizes the major criteria that supermarket shoppers in Ft. Collins, Colorado, considered important in selecting protein sources for meals (Roeber et al., 2002). With regard to beef purchases, the most important factor in purchasing was flavor and palatability followed by convenience, nutrient value, and price. Interestingly, these data would suggest that consumers trust the beef production system because food safety was ranked as the least important factor when buying protein, regardless of the source.

Consumer research in the 1980s and 1990s sent signals to the industry that the public had concerns about the product, which included

TABLE 1.1. Least squares mean rank (1 = most important, 6 = least important) for traits considered by consumers when buying protein sources for meals.

Protein source	Flavor/ palatability	Convenience	Health reasons	Nutrient value	Price	Safety
Beef	$1.40^{a,v}$	$2.76^{a,w}$	$4.36^{e,y}$	$2.95^{b,w}$	$3.75^{b,x}$	$5.12^{a,z}$
Pork	$1.43^{a,v}$	$2.91^{a,w}$	$4.00^{d,y}$	$3.40^{c,x}$	$3.20^{a,x}$	$5.33^{b,z}$
Poultry	$2.04^{b,w}$	$3.30^{b,xy}$	$3.03^{c,x}$	$3.39^{c,y}$	$3.08^{a,xy}$	$5.57^{c,z}$
Fish	$2.30^{c,v}$	$3.77^{c,x}$	$2.20^{a,v}$	$2.59^{a,w}$	$4.30^{c,y}$	$5.23^{ab,z}$
Nonmeat	$3.03^{d,y}$	$3.16^{b,y}$	$2.58^{b,x}$	$3.10^{b,y}$	$3.25^{a,y}$	$5.06^{a,z}$

Source: Adapted from Roeber et al. (2002).

^{a,b,c,d,e} Means, within a column, bearing a common superscript are not different ($P > 0.05$).

^{v,w,x,y,z} Means, within a row, bearing a common superscript are not different ($P > 0.05$).

tenderness, health, safety, price, and convenience. These concerns are now being addressed by targeted research projects (Ritchie and Corah, 1996).

Food safety and quality assurance became important societal concerns for the beef cattle industry during the early 1990s. The words *Escherichia coli* became a prominent part of all beef producers' vocabulary as cattle were identified as a potential reservoir of *E. coli* O157:H7 (Younts et al., 1999). Numerous *E. coli* O157:H7 outbreaks throughout the United States heightened the fear of the health risk posed by various food commodities, including ground beef. This led to the formation of numerous quality-assurance programs, the conducting of quality-assurance surveys, and a much more conscious awareness of sound quality-assurance practices by beef producers.

When the Food Marketing Institute (FMI) asked executives in the food service industry what single factor could most affect their business, their number one response was "food safety" (FMI, 2000). According to the FMI's "Trends in the United States" (2003), when consumers were questioned about the importance of a variety of factors that they considered when food shopping, a clean, neat store ranked first, followed by high-quality fruits and vegetables. In that survey, 79 percent of shoppers indicated that they were completely or mostly confident in the safety of foods in the supermarket. But consumers did believe that food safety problems could occur at various points in the food chain between the farm and their home. When asked where food safety problems were most likely to occur, consumers listed processing/manufacturing plants (35 percent), restaurants (15 percent), homes (10 percent), transportation (8 percent), grocery stores (6 percent), all of the above (18 percent), and farms (4 percent) (CAST, 2004).

A continuing trend for beef producers has been the documented change from just implementing quality-assurance/good-management practices to the verifiable documentation of these practices for the calves at the time of marketing (Corah, 2002). Fortunately for cattle producers, the majority of consumers presently perceive beef as a safe and wholesome product. However, there probably is no such thing as "too" safe when it comes to the food consumers who buy for themselves—and their children. Add to this the reality of the ever-increasing competition for the consumer's protein dollar, and you quickly see how crucial it is for cattle producers of all sizes in every segment to commit to a management strategy that inspires consumer confidence in the safety of beef products.

COMMON FOODBORNE DISEASES

Consumption of contaminated food has been estimated to cause 76 million illnesses, 325,000 hospitalizations, and 5,200 deaths in the United States each year (Centers for Disease Control and Prevention, 2000, available at: www.cdc.gov/ncidod/dbmd/diseaseinfo/foodborneinfections_t .htm). The most commonly recognized foodborne infections are those caused by the bacteria *Campylobacter, Salmonella,* and *E. coli* O157:H7, and by a group of viruses called calicivirus, also known as the Norwalk and Norwalk-like viruses.

More than 250 foodborne diseases have been described. The symptoms of foodborne infections vary widely, with diarrhea and vomiting being the most common. Medical costs and lost wages due to foodborne salmonellosis, which is only one of many foodborne infections, have been estimated to be more than $1 billion each year. Interestingly the proportion of outbreaks caused by fruits and vegetables consumed without cooking has increased over recent decades.

The following is a brief description of these common foodborne diseases.

Escherichia coli *O157:H7*

Escherichia coli O157:H7 is a strain of *E. coli* producing toxins (or poisons) that can damage the lining of the human intestine. *E. coli* O157:H7 infection can magnify quickly in humans, particularly in children, seniors, or people who have weak immune systems. For these individuals, *E. coli* O157:H7 can become life-threatening by causing kidney failure.

E. coli O157:H7 can be found throughout the environment; any place where animals (including pets) coexist with humans. *E. coli* O157:H7 survives remarkably well, and when an animal has *E. coli* O157:H7 in its intestine, it typically "sheds" the organism through its feces. Beef carcasses can become contaminated with this *E. coli* strain during the harvesting process, such as when the hide is removed from a carcass and it inadvertently touches the carcass surface. *E. coli* O157:H7 has been found in sheep, cattle, horses, goats, elk, pigs, deer, opossums, raccoons, dogs, poultry, wild birds, and houseflies. The organism has been found in young beef calves and older cows, in dairy calves, and in dairy cows. Cattle coming into the feedlot carry the organism, as do hides within processing facilities (BIFSCo, available at: http://www.bifsco.org/udocs/E.%20Coli%20a%20Basic%20Look.pdf).

E. coli O157:H7 is difficult to control because it adapts and survives in many different environments. The organism can remain viable for months at a time in both feces and soil. In the past, the beef industry suffered a number of food safety setbacks, primarily due to *E. coli* O157:H7 contamination of ground beef. Fortunately, beef processors became very proactive in implementing postharvest strategies to reduce the risk of foodborne illnesses (organic acid rinses, steam vacuuming, steam pasteurization, etc.).

Campylobacter jejuni

Campylobacter jejuni is the leading cause of bacterial diarrhea illness in the United States, and causes more disease than the *Shigella* spp. and *Salmonella* spp. pathogens combined. Campylobacteriosis is the name of the illness caused by *C. jejuni*. It is also known as campylobacter enteritis or gastroenteritis. Although *C. jejuni* is not carried by healthy individuals in the United States or Europe, it is often isolated from healthy cattle, chickens, birds, and even flies. This pathogen requires reduced oxygen levels to survive, and is relatively fragile and sensitive to environmental stresses such as drying, heating, disinfectants, and acidic conditions.

C. jejuni frequently contaminates raw chicken. Surveys show that 20 to 100 percent of retail chickens are contaminated, which is not surprising since many healthy chickens carry *C. jejuni* in their intestinal tracts. Raw milk can also be a source of infection. The bacteria are often carried by healthy cattle and by flies on farms. Non-chlorinated water may also be a source of infections. However, cooking chicken or meat properly, pasteurizing milk, and chlorinating drinking water will kill the bacteria. Research has identified the top risk factors for campylobacteriosis as handling raw poultry and eating undercooked poultry.

Salmonella

It is estimated that between 2 and 4 million cases of salmonellosis occur in the United States annually. *Salmonella* cycle continuously through the environment via the intestinal tracts of animals and humans. *Salmonella* bacteria are most commonly found in raw or undercooked foods such as poultry, meat, eggs, unpasteurized milk, or other dairy products. *Salmonella* infections are hard to trace, because those who have the symptoms often do not associate them with food. Symptoms can include headache, abdominal pain, diarrhea, fever, and nausea, and generally begin six to 48 hours after eating contaminated food. As with other bacteria, proper cooking and handling can dramatically reduce the incidence of salmonellosis. Compared to other meats and poultry, the incidence of *Salmonella* organisms on fresh beef is quite low.

Listeria monocytogenes

Foods contaminated with the bacterium *Listeria monocytogenes* can cause listeriosis, an infection that can seriously affect pregnant women, newborns, the elderly, and adults with weakened immune systems. According to the Centers for Disease Control (CDC), in the United States, an estimated 2,500 persons become seriously ill with listeriosis each year.

Listeria monocytogenes is a common environmental bacteria found in soil and water, as well as in a variety of raw foods such as uncooked meats, dairy products, and vegetables. Proper cooking, storage, handling, and preparation of foods significantly reduce problems with *Listeria* in foods.

Yersinia enterocolitica

Strains of *Y. enterocolitica* can be found in meats (pork, beef, lamb, etc.), oysters, fish, and raw milk. The exact cause of the food contamination is unknown. However, the prevalence of this organism in the soil and water and in animals, such as beavers, pigs, and squirrels, offers ample opportunities for human contact.

Yersiniosis is frequently characterized by gastroenteritis with diarrhea and/or vomiting; however, fever and abdominal pain are the hallmark symptoms. *Yersinia* infections often mimic appendicitis.

Table 1.2 summarizes incidence and death rate for states monitored by FoodNet (2003) for the Centers for Disease Prevention.

TABLE 1.2. Incidence and death rate by organism.

	Cases		Deaths	
Organism	**No.**	**Rate**[a]	**No.**	**Rate**[b]
Salmonella	6,043	14.43	34	0.68
Campylobacter	5,273	12.60	9	0.22
Shigella	3,041	7.27	2	0.08
Cryptosporidium	481	1.09	3	0.68
E. coli	444	1.06	4	0.94
Yersinia	162	0.39	2	1.53
Listeria	139	0.33	22	16.54
Vibrio	110	0.26	7	7.69

Source: FoodNet (2003).

[a]Cases per 100,000 population for FoodNet areas.

[b]Deaths per 100 cases with known outcome.

These results show that *Salmonella, Campylobacter,* and *Shigella* would be the main sources of foodborne illness. The rate of infection for *Salmonella* was approximately 14 times higher than for *E. coli.*

McMullen (2001) summarized the incidence of foodborne pathogens measured on the carcasses of beef, pork, broilers, and turkeys using USDA data from 1994 to 1998 (Table 1.3). These results revealed that the highest concentrations of *E. coli* were on beef carcasses (0.2 percent) and were nondetectable for pork, broiler, and turkey carcasses. However, the highest incidences of *Salmonella* were found in broilers and turkey carcasses (~19 percent) and lowest for beef carcasses (1 percent).

Data summarized by the FoodNet sites from five U.S. states suggested that the incidence of *E. coli* O157:H7 infections is continuing to decline (Table 1.4). The incidences have dropped from 2.8 cases/100,000 in 1998 to 1.7 cases/100,000 in 2002.

Foodborne pathogens remain a primary focus for the beef industry and its consumers, but significant progress has been accomplished. From 1996 to 2004, the incidence of *E. coli* O157:H7 infections decreased by 42 percent. *Campylobacter* infections fell by 31 percent, cryptosporidium dropped by 40 percent, and *Yersinia* decreased by 45 percent. The Centers for Disease Control and Prevention recently announced that the overall incidence of *E. coli* O157:H7 infections declined by 9.5 percent in 2004 alone. In addition, the USDA recently reported that the percentage of ground beef samples testing

TABLE 1.3. Prevalence of selected foodborne pathogens on beef, pork, and poultry carcasses.

	Percentage of positive carcasses			
Foodborne pathogens	**Beef**	**Pork**	**Broiler carcass**	**Turkey carcass**
Clostridium perfrigens	2.6	10.4	42.9	29.2
Staphlococcus aureus	4.2	16.0	64.0	66.7
Listeria monocytogenes	4.1	7.4	15.0	5.9
Campylobacter jejuni/coli	4.0	31.5	88.2	90.3
Escherichia coli	0.2	0.0	0.0	0.0
Salmonella	1.0	8.7	20.0	18.6

Source: McMullen (2001). Used by permission.

TABLE 1.4. Incidence of *Escherichia coli* O157:H7 infections based on five original FoodNet sites in the United States.

Year	CA	CT	GA	MN	OR	Overall incidence
	No. of cases per 100,000					
1996	1.1	2.3	0.5	5.2	2.3	2.7
1997	0.9	1.4	0.2	4.2	2.5	2.3
1998	NDR[a]	NDR	NDR	NDR	NDR	2.8
1999	NDR	NDR	0.6	NDR	NDR	2.1
2000	NDR	NDR	0.5	4.4	NDR	2.7 (2.0[b])
2001	1.1	1.1	0.6	4.8	2.3	(1.6[b])
2002	1.0	1.4	0.7	3.6	5.1	(1.7[b])

Source: Council for Agricultural Science and Technology (CAST) (2004). Intervention strategies for the microbiological safety of foods of animal origin. Issue paper 25. CAST, Ames, Iowa. Used by permission.

[a]No data reported.

[b]Nine sites.

positive for *E. coli* O157:H7 declined by more than 80 percent since 2000, including a 43.3 percent year-over-year reduction between 2003 and 2004. "I think that alone illustrates how far we've come," said Jim McAdams, president of the National Cattlemen's Beef Association (BIFSCo, 2005). West noted that these results already exceed the national goal of reducing incidences of *E. coli* O157:H7 to 1.0 per 100,000 people by 2010. Results presented in Table 1.5 also support these observations. The prevalence of *E. coli* O157:H7 appeared to be declining since 2000 (down from 0.86 percent positive to 0.36 percent positive).

WHY FOODBORNE PATHOGENS ARE OF CONCERN TO THE BEEF INDUSTRY

Table 1.6 indicates how much investment was required to reduce the incidence of *E. coli* O157:H7 in beef carcasses using postharvest and also how *E. coli* negatively influenced buying decisions of the consumer. Steve Kay estimated that during the period from 1993 to 2002, *E. coli* O157:H7 cost the beef industry approximately $2.7 billion dollars (Table 1.6).

TABLE 1.5. Prevalence of *Escherichia coli* O157:H7 in ground beef (CAST, 2004).

Year	No. of samples	No. of positives	% Positive
1995[a]	5,291	3	0.057
1996[a]	5,326	4	0.075
1997[a]	5,919	2	0.034
1998[b]	7,529	14	0.19
1999[b]	8,710	29	0.33
2000[c]	6,374	55	0.86
2001[c]	7,009	59	0.84
2002[c]	7,026	55	0.78
2003[c]	~5,103	18	0.35

Source: Council for Agricultural Science and Technology (CAST) (2004). Intervention strategies for the microbiological safety of foods of animal origin. Issue paper 25. CAST, Ames, Iowa. Used by permission.

[a]25-g analytical samples.

[b]325-g analytical samples.

[c]325-g analytical samples and improved isolation method (i.e., immunomagnetic beads to concentrate cells).

TABLE 1.6. The costs of reducing *E. coli* contamination in beef from 1993 to 2002.

Item	Cost in millions of dollars
Impact on demand	$1,584
Impact on boneless beef prices	$172
Capital expenditures by top 10 beef packers	$400
Increased operating costs for top 10 beef packers	$250
Spending by next 20 beef packers	$100
Retail costs incurred by packers	$100
Government and industry research	$65
Total Costs	$2,671

Source: Adapted from Kay (2003).

In order to continue to focus efforts on continued improvements in beef safety, the Beef Industry Food Safety Council (BIFSCo) was formed in October 1997 to develop industry-wide, science-based strategies to solve the problem of *E. coli* O157:H7 and other foodborne pathogens in beef. Bovine spongiform encephalopathy (BSE, or mad cow disease) is another top priority for BIFSCo and its research efforts.

BIFSCo focuses on prevention at all stages of the beef production process to significantly reduce and potentially eliminate foodborne illnesses in beef, and this is accomplished through the following objectives (available at: www.beef.org/ncbasafety.aspx#bifsco):

- Identify, prioritize, and facilitate research activities from farm to table.
- Develop programs to help industry segments operate in today's business environment.
- Speak with one voice in seeking regulatory and legislative solutions.
- Prepare consumer education programs.
- Initiate and implement industry education programs.

There continues to be a dedicated effort toward understanding the association between *E. coli* O157:H7 concentration in live cattle and different management practices. While postharvest multiple-hurdle methodologies have been effective in reducing *E. coli* O157:H7 contamination of carcasses, less research has been directed at preharvest strategies. In a recent summary, Sargeant et al. (2004) outlined the factors that were associated with *E. coli* O157:H7 in cattle feces. Among the factors determined were (1) the frequency of observing cats in the pens of alleys; (2) the presence of *E. coli* O157:H7 in water tanks; (3) the use of antibiotics in the ration or water (negative association; the wetness of the pen, number of cattle in the pen); and (4) wind velocity and height of the feed bunk.

Foodborne diseases are largely preventable, though there is no simple one-step prevention measure such as a vaccine. Instead, measures are needed to prevent or limit contamination all the way from farm to table. A variety of good agricultural and manufacturing practices can reduce the spread of microbes among animals and prevent the contamination of foods. Careful review of the whole food production process can identify the principal hazards, and the control points where contamination can be prevented, limited, or eliminated. A formal method for evaluating the control of risk in foods exists and is called the Hazard Analysis and Critical Control Point, or HACCP system (Texas Beef Quality Assurance Manual, 2005, available at: www.texasbeefquality.com).

HACCP safety principles are now being applied to an increasing spectrum of foods, including meat, poultry, and seafood. It is a process of identifying what could go wrong, of making plans to avoid it, and of documenting what one has done. The HACCP principles involve seven steps:

1. *Analyze hazards.* Potential hazards associated with a food and measures to control these hazards are identified. The hazard could be biological, such as a microbe; chemical, such as a toxin; or physical, such as ground glass or metal fragments.
2. *Identify critical control points.* These are points in a food's production—from its raw state through processing and shipping to consumption by the consumer—at which the potential hazard can be controlled or eliminated. Examples are cooking, cooling, packaging, and metal detection.
3. *Establish preventive measures with critical limits for each control point.* For a cooked food, for example, this might include setting the minimum cooking temperature and time required to ensure the elimination of any harmful microbes.
4. *Establish procedures to monitor the critical control points.* Such procedures might include determining how and by whom cooking time and temperature should be monitored.
5. *Establish corrective actions to be taken when monitoring shows that a critical limit has not been met.* For example, reprocess or dispose of food if the minimum cooking temperature is not met.
6. *Establish procedures to verify that the system is working properly.* For example, test time-and-temperature recording devices to verify that a cooking unit is working properly.
7. *Establish effective record-keeping to document the HACCP system.* This would include records of hazards and their control methods, the monitoring of safety requirements and action taken to correct potential problems. Each of these principles must be backed by sound scientific knowledge: for example, published microbiological studies on time-and-temperature factors for controlling foodborne pathogens (http://www.cfsan.fda.gov/~lrd/bghaccp.html).

Table 1.7 illustrates how implementation of HACCP principles was effective in reducing the incidence of *Salmonella* on carcasses of different animals. In every case, after following HACCP implementation, there was reduced incidence of *Salmonella*.

SUMMARY

The Council of Agriculture and Science and Technology (CAST, 2004) outlined numerous suggestions for improving the safety of meat in the United States. Among the twelve recommendations that were suggested were the following:

(1) Improving the safety of foods of animal origin begins at the farm.
(2) Innovative intervention strategies are needed to control pathogens in animal production.
(3) New intervention strategies that decrease public health hazards—strategies such as on-farm treatments, antimicrobial treatments, food additives, and processing technologies—should receive expedited review by regulatory agencies.
(4) Good agricultural practices are needed that effectively decrease pathogens during production and include effective intervention strategies for beef, dairy, pork, egg, and poultry producers.
(5) On-farm food safety practices should be developed and implemented based on management programs used in food-manufacturing facilities.
(6) Practices on the farm used to produce milk and meat should reflect the latest science in animal health and well-being, public health, environmental health, and medical ecology and should contribute to an economically viable industry.
(7) All sectors of agriculture must have sufficient funding to refine management practices and to train producers, veterinarians, and allied industry personnel in leading-edge management techniques.

If the beef industry is going to continue to be successful, Chuck Schroeder (2003) presented the following three rules for the beef industry to adhere to:

TABLE 1.7. Prevalence of *Salmonella* before and after implementation of HACCP procedures.

Product	Incidence before HACCP	Incidence after HACCP
Broilers	20.0	10.9
Swine	8.7	6.5
Steers and heifers	1.0	0.2
Cows and bulls	2.7	2.2
Ground beef	7.5	4.8
Ground turkey	49.9	36.4

Source: McMullen (2001). Used by permission.

Rule 1. Remember you are in a consumer product business. Everyone has an obligation to deliver quality, affordability, and an overall image ("satisfaction") that will cause the consumer to choose beef over its competitors in the meat case or on the menu. Each one in the system must be accountable for his or her contribution to meeting the consumer's desires, and should expect to be compensated for that contribution.

Rule 2. Remember you are in a changing business. There is scarcely a successful consumer product on the market today that is the same as it was five years ago—in some cases, five months ago! The complex beef industry requires agility to respond to the many factors influencing its potential for profitability, such as: consumer demands for innovation; rising costs of land, labor, and other inputs; regulations driven by fact, fear, and perceptions; competition from other proteins; dwindling sources of labor, etc. We resist change for a variety of good and poor reasons, but those who grasp it and drive it have a chance to turn it to their advantage. We must strategically invest available industry resources, in understanding, preparing for, and effecting proper change.

Rule 3. Remember you are in a relationship business. While the independent nature of beef producers, processors, and marketers has been a signature of our culture, everyone must recognize the concomitant nature of business relationships today. Sparking conflict for the sake of tradition and entertainment is a costly practice that benefits only our competitors, and contributes to the outward squeeze on human resources as it adds cost to the system. Learn to know and respect your customers. Learn to know and respect your suppliers. Learn to know and respect your peers. If the regular reading, viewing, and listening materials on your industry do not contribute to that knowledge and respect, stop reading, viewing, and listening to them. The beef industry has turned an important corner in recent years as traditional lip service to consumer needs was transformed into strategic action. That action resulted in a measurable change in consumer demand, which translated into more consumer dollars flowing into the beef system. If that path will be continued to be followed, in spite of the often self-generated distractions and the turns it will invariably take, it will lead to an industry that is growing on sound fundamentals, attracting investment and entrepreneurial vigor, and establishing new strength on its traditional values.

LITERATURE CITED

BIFSCo. 2005. Executive Summary of the 2005 Beef Industry Safety Summit. National Cattlemen's Beef Association, Centennial, CO, p. 1.

CAST. 2004. Council of Agriculture, Science and Technology. Intervention strategies for the microbiological safety of foods of animal origin. Issue paper 25.

Centers for Disease Control and Prevention. 2000. Surveillance for Foodborne-Disease Outbreaks—United States, 1993–1997. U.S. Department of Health and Human Services, Atlanta, GA. Available at: ftp.cdc.gov/pub/Publications/mmwr/ss/ss4901.pdf.

Corah, L.R. 2002. How to achieve profitability in a dynamic beef industry. Int'l. Livestock Congress, Houston, TX. Available at: www.theisef.com/component/option,com_docman/task,cat_view/gid,70/Itemid,92/.

FMI. 2000. The food marketing industry speaks. Food Marketing Institute, Washington, DC.

FMI. 2003. Trends in the United States, 2003. *Consumer Attitudes and the Supermarket.* Food Marketing Institute, Washington, DC.

FoodNet. 2003. Foodborne diseases active surveillance network (FoodNet) emerging infections program report on foodborne pathogens. Available at: www.cdc.gov/foodnet/pub/publications/2005/FNsurv2003.pdf.

Genho, P.C. 2003. The producer's role in producing consumer-demanded beef. Int'l Livestock Congress, Houston, TX. Available at: www.theisef.com/component/option,com_docman/task,cat_view/gid,70/limit,5/limitstart,10.

Kay, S. 2003. $2.7 billion: The cost of *E. coli* O157:H7. *Meat and Poultry* (February), pp. 26-34.

McCollum, T. 2003. The impact of health on performance, profits and carcass quality on value added calves. Presented at the Rocky Mountain Beef Options Seminar, Billings, MT, pp. 51-53.

McMullen. L.M. 2001. Bacterial contamination of meat products—how serious is the problem and what can the livestock industry do to be proactive? Proceedings of the 22nd Western Nutrition Conference, Saskatoon, Saskatchewan, September 21-22.

Oldham, C. 2002. Adding value . . . keeping customers. Available at www.aginfolink.com.

Paterson, J. 2002. Surviving and thriving in the next decade. Proceedings of the International Livestock Congress, Houston, TX, pp. 25-34.

Ritchie, H.D. and L.R. Corah. 1996. Beef cattle research: Where have we been and where are we headed? Prepared for NCA Research and Education Committee, January 29, San Antonio, TX.

Roeber, D.L., J.A. Scanga, K.E. Belk, and G.C. Smith. 2002. Consumer attitudes and preferences. Animal Sciences Research Report. The Department of Animal Sciences, Colorado State University, Fort Collins.

Sargeant, J.M., M.W. Sanderson, R.A. Smith, and D.D. Griffin. 2004. Associations between management, climate and *Escherichia coli* O157 in the feces of feedlot cattle in the Midwestern USA. *Prev. Vet. Med.* 66 (1-4): 175-206.

Schroeder, C. 2003. The issues in profitably producing consumer-demanded beef. Proceedings of the International Livestock Congress, March, Houston, TX, pp 1-5.

Smith, G.C. 2001a. Creating opportunities in a beef supply chain. Presented at the Mississippi Cattlemen's Association Annual Convention, Jackson, MI.

Smith, G.C. 2001b. Increasing value in the beef supply chain. Presented for Boehringer/Ingelheim (Canada) Ltd., Vetmedica Division, in Coaldale, Lethbridge and Kananaskis, Alberta, June.

Smith, G.C. 2003. Beef production and marketing in the 21st century. Presented at the annual meeting of the Canadian Cattlemen's Association, Moose Jaw, Saskatchewan, August.

Texas Beef Quality Assurance. 2005. Available at: www.texasbeefquality.com/.

Younts, S.M., E.C. Alocilja, W.H. Osburn, S. Marquie, and D.L. Grooms. 1999. Development of electronic nose technology as a diagnostic tool in the detection and differentiation of *Escherichia coli* O157:H7. *J. Anim. Sci.* 77 (Suppl. 1): 129.

Chapter 2

Preharvest Beef Safety: Production Management and Pathogen Control

Guy H. Loneragan
Mindy M. Brashears

PREHARVEST CONTROL OF PATHOGENS: IS IT POSSIBLE AND SHOULD IT EVEN BE CONSIDERED?

Production of a consistently safe and wholesome product is essential for continued consumer confidence and loyalty to beef. Events, either perceived or real, that raise questions concerning the safety of beef or the commitment of beef producers to protect consumers' health serve to highlight the intimate link between demand for beef and producer profitability. Beef maintains center-of-the-plate status due in part to consumers' confidence in its quality, consistency, and safety. Despite the overall excellent microbial safety of beef and the commitment of all segments of the beef industry to produce a safe product, various human pathogens that frequently inhabit the bovine gastrointestinal tract provide challenges that must be overcome.

Much effort in the safety of the product has targeted harvest-level controls. This is the point at which most reward for investment can be achieved by preventing contamination of carcass surfaces or removing it once it has occurred. Implementation of efficacious interventions during harvest also makes practical sense because all cattle intended

Handbook of Beef Safety and Quality

doi:10.1300/5640_02

for human consumption must pass through packing plants regardless of preharvest production practices (such as feedlot, dairy, grass-fed, etc.) Moreover, microbial interventions are generally non-species-specific and are stable across regions. In other words, the effect of steam pasteurization on *E. coli* O157 should not vary greatly depending on region, seasons, or on whether the bacterium originated from a cull dairy cow or a feedlot steer.

That said, however, there has been increasing interest in preharvest control, particularly of *E. coli* O157 (Brashears and Loneragan, 2005), the theory being that if carriage by cattle entering plants can be reduced, then the likelihood of carcass contamination is also reduced (Arthur et al., 2004; Elder et al., 2000). As a consequence, preharvest control would serve as an important hurdle in a multihurdle approach and ultimately improve the efficiency of harvest-level interventions. For this conceptual approach to work in practice (1) it must be possible to alter carriage of pathogens by cattle and (2) preharvest status must be associated with postharvest contamination. This latter association need not be at the animal level but must at least be demonstrated at the pen or group level.

Recent data has enabled us to evaluate the latter proviso in part because *E. coli* O157 is ubiquitous among cattle populations yet its pen-level burden varies greatly within a farm (such as a feedlot) and across seasons (Barkocy-Gallagher et al., 2003; Hancock, Besser, Rice, Herriott, and Tarr, 1997; Hancock, Rice, Thomas, Dargatz, and Besser, 1997; Loneragan and Brashears, 2005a; Sargeant, Sanderson, Smith, and Griffin, 2004; Smith et al., 2005). As a consequence of pen-to-pen variation, it has been possible to evaluate pen-level carriage (prevalence) at harvest and subsequent prevalence in the group-specific carcasses. Elder and co-workers (Elder et al., 2000) recovered *E. coli* O157 from 26.2, 13.0, 43.4, 18.3, and 1.9 percent of fecal samples, hides, pre-evisceration carcasses, post-evisceration carcasses, and carcasses post-intervention, respectively. In their study, preharvest prevalence (i.e., carriage in either feces or on hides) was associated ($r = 0.58$, 95 percent, CI 0.27-0.78, $P < 0.01$) with recovery of the organisms from the group-specific carcasses (at any site either pre- or post-evisceration, or post-intervention). In other words, if the pen-level carriage of *E. coli* O157 increased, so too on average did the recovery from carcasses. Importantly, most carcass contamination occurred early in production, indicating that hides may be the major source. In a separate study, Ransom and colleagues collected fecal samples from feedlot pens of cattle prior to their shipment to a packing plant (Ransom et al., 2003). Within the plant, they sampled hides and carcasses from these pens of cattle. They recovered *E. coli* O157 from 22.5, 46.3, 12.5, 2.5, and 0.6 percent of hides, colons, pre-evisceration carcasses, post-evisceration carcasses, and carcasses in the cooler, respectively. The likelihood of pathogen recovery in the plant was associated with pen-level fecal prevalence at the feedlot; prevalence was collapsed into two categories, that is, *greater than 20 percent* or *20 percent or less*. In pens categorized as the latter, *E. coli* O157 was recovered from 5.7, 7.1, 7.1, 0.0, and 0.0 percent of hides, colons, pre-evisceration carcasses, post-evisceration carcasses, and carcasses in the cooler, respectively. In pens with a fecal prevalence greater than 20 percent, *E. coli* O157 was recovered from 22.5, 46.3, 12.5, 2.5, and 0.6 percent of hides, colons, pre-evisceration carcasses, post-evisceration carcasses, and carcasses in the cooler, respectively. More recently, *E. coli* O157 was recovered from 75.7 percent of hides, 14.7 percent of pre-evisceration carcasses, 3.8 percent of post-evisceration carcasses, 0.3 percent of post-intervention samples, and 0.0 percent of chilled carcasses (Arthur et al., 2004). The authors of this study reported that *E. coli* O157 prevalence on hides was associated ($R^2 = 0.68$, $P < 0.05$) with the prevalence on pre-evisceration carcasses. These data, and those of others, indicate that hides are the primary source of contaminants on carcasses.

These studies demonstrate that contamination of carcasses with *E. coli* O157 is proportional to the group-level burden entering plants. Despite the efficacious in-plant interventions, if the pathogen-load on hides (as estimated by prevalence) entering the plant is sufficiently large then the likelihood of recovering *E. coli* O157 from the carcass contamination (including post-intervention) also increases. While not certain, it is probable that the pre- and postharvest relationship observed for *E. coli* O157 is not primarily determined to be a unique characteristic of that organism. In other words, the relationship observed should be stable across a variety of pathogens (Arthur et al., 2004; Bosilevac et al., 2004). If, therefore, it is possible to manipulate carriage of pathogens in the preharvest environment, then, presumably, it is possible that efficacious preharvest interventions will reduce human exposure and the public health burden of foodborne pathogens where beef is implicated as a vehicle for exposure.

PATHOGENS OF INTEREST: PAST AND FUTURE

Over the past two decades, the attention of consumers, regulators, and producers has largely been focused on shiga toxinogenic *E. coli* O157 in beef products, in particular, ground beef (Anonymous, 1994, 1997, 2002a; Grimm et al., 1995; Jackson et al., 2000; Kassenborg et al., 2004; Laine et al., 2005; Ostroff et al., 1990). Great strides have been made in reducing its presence in beef products (Figure 2.1) and the occurrence of human disease (Figure 2.2). Despite substantive successes, more progress can and should be achieved, particularly in the arena of preharvest control. Preharvest interventions need to be developed and implemented for a variety of different production systems, such as confined feeding and grass-fed beef. Not only will im-

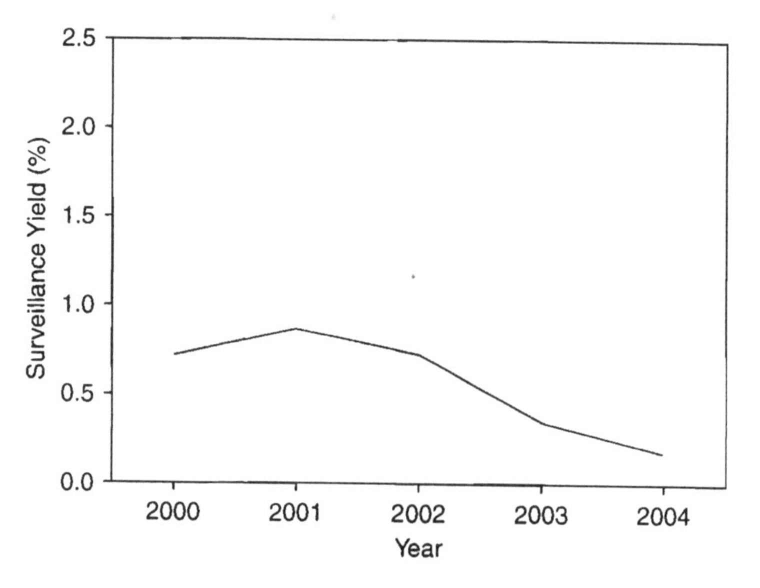

FIGURE 2.1. United States Department of Agriculture, Food Safety, and Inspection Service (FSIS) reported ground beef samples positive for *E. coli* O157 through random sampling. Samples are tested. *Source:* www.fsis.usda.gov/OPHS/ecoltest/tables1.htm. Accessed January 23, 2006.

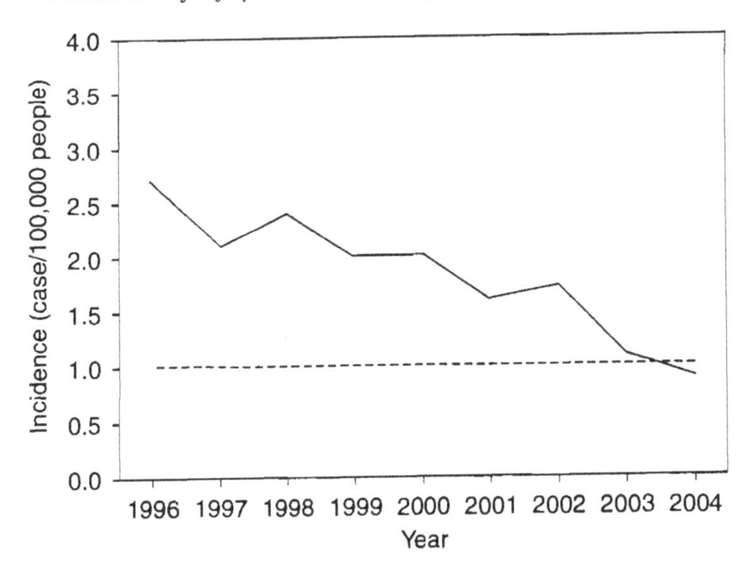

FIGURE 2.2. Centers for Disease Control and Prevention (CDC) estimate for human incidence of *E. coli* O157-induced illnesses (solid line) based on FoodNet data. Incidence (Y axis) is reported in cases per 100,000 people in reporting states. (*Source:* www.cdc.gov/foodnet/reports.htm. Accessed January 23, 2006.) Dashed line represents Healthy People 2010 Objective. Source: www.healthy people.gov/Publications/. Accessed January 23, 2006.

plementation of a systematic suit of effective interventions reduce carcass (and beef) contamination with *E. coli* O157 but it should also reduce environmental contamination and consequently, human exposure through non-beef sources.

Other challenges that are being brought into sharper focus include non-O157 shiga toxinogenic *E. coli* (STEC) variants (Hussein and Bollinger, 2005; Renter et al., 2005), *Salmonella* (particularly those carrying resistance determinants to important human-use antimicrobial drugs (Anonymous, 2002b; Fey et al., 2000; Gupta et al., 2003), and antimicrobial drug resistance in general (Angulo et al., 2004). Non-O157 STEC is the predominant STEC associated with cattle or human disease in some countries (Hussein and Bollinger, 2005). Regional factors that favor a particular STEC variant over another are not certain and, just as importantly, it is unclear how stable serotype predominance is over time. In other words, just as shifts in *Salmonella enterica* serovars occur over time, it is possible that *E. coli* O157 may not always be the predominant STEC associated with either cattle or human illness in the United States in the years to come. If this is so, a better understanding of non-O157 STECs and strategies to evaluate temporal changes are needed. Estimates of prevalence of non-O157 STEC are 4.6 to 10.0 for *E. coli* O26 and 0.0 to 8.0 for *E. coli* O111 (Pearce et al., 2006; Renter et al., 2005; Loneragan, unpublished data, 2006).

Antimicrobial drug-resistant *Salmonellae,* such as *Salmonella* Typhimurium DT104 and *Salmonella* Newport MDR-AmpC are important foodborne pathogens that are particularly problematic because they are resistant to multiple antimicrobial drugs (Anonymous, 2002b; Helms et al., 2005). In some cases, *Salmonella* are resistant to primary drugs of choice (Fey et al., 2000; Gupta et al., 2003; Helms et al., 2002). When antimicrobial therapy is indicated, therefore, drug choice is limited to second-line products. Moreover, some authors infer that disease associated with resistant variants is more severe than disease associated with pansusceptible isolates (Devasia et al., 2005; Helms et al., 2002; Varma et al., 2005). It is probable that *Salmonella* do make their way from animals to humans through food products. What is controversial, however, is the role of antimicrobial use in food animal production systems, selection of resistant determinants, and disease in humans due to resistant variants (Angulo et al., 2004; Fey et al., 2000; Hurd et al., 2004; Phillips et al., 2004).

Dealing with antimicrobial resistance may prove a far more complex issue than controlling a specific pathogen. While it has been repeatedly demonstrated that antimicrobial drug use exerts a selection pressure that favors bacteria that are resistant to that antimicrobial, modifiable factors that influence acquisition, dissemination, and maintenance of resistance are far from certain. In addition, whether resistance determinants are housed on the bacteria or plasmid chromosome may be an important consideration. In other words, control strategies might be directed at the specific genera of bacteria most likely to harbor resistance determinants or at the resistance determinants themselves independent of specific genera. Some countries have chosen to arbitrarily ban the use of some or all in-feed antimicrobial drugs. It may be possible to develop specific interventions that effectively mitigate dissemination of genetic material encoding resistance without arbitrarily prohibiting use of certain drugs.

Regardless of the future challenges, the threat posed by and response to *E. coli* O157 clearly demonstrate that it is possible to develop critical control points at which interventions can be deployed with great efficacy. This is not meant to imply that *E. coli* O157 is no longer a public health burden but rather that the development and implementation of interventions, which is an ongoing process, has produced significant beneficial results. Some of these interventions are likely effective against most bacteria whereas others have a more limited spectrum of efficacy. Challenges associated with other pathogens will benefit from the discovery experience of developing interventions for *E. coli* O157.

The focus of this discussion of interventions will be limited to those interventions developed for use against *E. coli* O157 in the preharvest environment. There are two broad conceptual opportunities for preharvest control of *E. coli* O157. The first conceptual avenue for control is modification of an existing management strategy such as manipulation of a feed ingredient, environmental modification, or variation of the frequency with which water troughs are washed. The second avenue for control is development and implementation of specific intervention technologies. If the former avenue for control is proven effective, then it would be preferable because modification of existing management practices could be implemented expeditiously with minimal training of personnel. This is not to imply that the second avenue is not desirable but it would require significant investment of time and money in the areas of discovery, development, and implementation. Many benefits of *E. coli* O157 control are not immediately tangible or self-evident for beef producers. As such, the reluctance of some cattlemen should not be underestimated to bear the required costs associated with implementation of new technologies whose immediate benefits are not directly related to day-to-day cattle raising and feeding.

Studies to date have failed to identify management practices that are consistently associated with *E. coli* O157 prevalence and, importantly, that can readily be modified. An association with muddy pen surfaces and *E. coli* O157 prevalence was documented by Smith et al. (2001) but this variable was not consistent in another study (Sargeant,

Sanderson, Smith, and Griffin, 2004). In addition, presence in water troughs has been associated with carriage (LeJeune et al., 2001; Sargeant, Sanderson, Griffin, and Smith, 2004) yet water trough sanitation (physical or chemical) has shown little or no effect on prevalence (Elder and Keen, 1999; LeJeune et al., 2004). Season appears to be the most consistently predictable modifier of carriage (Anonymous, 2001; Barkocy-Gallagher et al., 2003; Hancock, Besser, Rice, Herriott, and Tarr, 1997; Smith et al., 2005); the reason for the seasonal influence on *E. coli* O157 shedding is uncertain and, unfortunately, while seasonal variation in shedding is repeatable, it is necessarily not modifiable. Until stable and modifiable factors can be identified, it is unlikely that it will be able to effectively control *E. coli* O157 through modification of existing management practices.

Based on our current knowledge of the epidemiology of *E. coli* O157 in cattle production environment, preharvest control will require development and implementation of preharvest interventions. The interventions discussed later are not inclusive of all potential/proposed technologies. Rather, information on the interventions that are currently available for use, that are pending regulatory approval, or that have had substantive studies performed in existing, appropriate settings is presented. Focus has been placed on interventions that appear to exert a positive impact. Interventions with marginal benefits such as chlorination of water (LeJeune et al., 2004; Loneragan, unpublished data, 2006) or *Ascophyllum nodosum* (brown seaweed) are not included. It is important to note that not all proposed interventions are efficacious.

Bacteria on hides appear to be the primary source for bacterial contamination of carcasses. Some argue that intervention studies in which carriage on hides was not evaluated carry less inferential weight than those studies in which the effect of the intervention of hide carriage was evaluated. We recently published data quantifying the relationship between pen-level prevalence in feces and on hide (Loneragan and Brashears, 2005b). Our modeling indicates that pen-level fecal prevalence is predictive of carriage on hides (Figure 2.3). Consequently, there are few real inferential limitations of studies in which feces were the sole measure of the intervention. In other words, the data demonstrate that an intervention effectively reduces fecal carriage; it will, on average, also reduce hide contamination and subse-

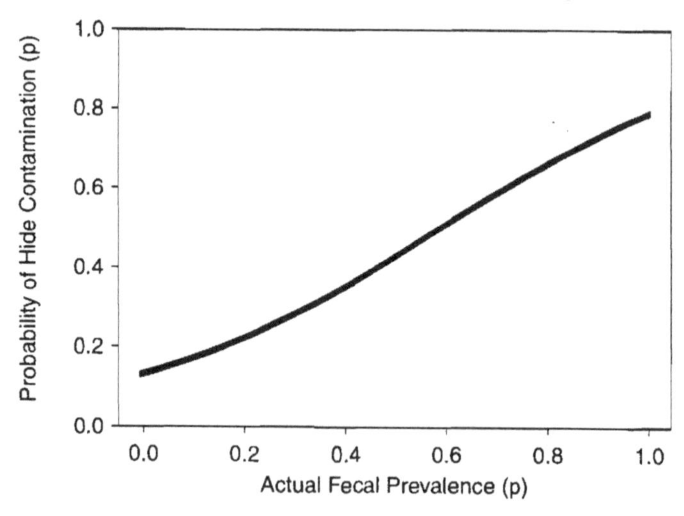

FIGURE 2.3. Hide prevalence as a function of prevalence (solid line). The shaded area represents the 95 percent confidence limits. *Source:* Adapted from Loneragan and Brashears (2005b).

quent likelihood of carcass contamination. Effective preharvest control of *E. coli* O157 could be achieved with an intervention directed at reducing hide contamination if applied just prior to harvest. Alternatively, long-term control could be effected with an intervention directed at reducing colonization at the rectoanal junction (Naylor et al., 2003) and other sites within the gastrointestinal tract.

DIRECT-FED MICROBIALS

This broad category includes probiotics and competitive exclusion. The concept of competitive exclusion is relatively new and was first effectively exploited in poultry (Callaway et al., 2003b). Competitive exclusion is based on the concept that nonpathogenic organisms (1) compete for limited nutrients or (2) produce products that are inhibitory to other microorganisms. Probiotics on the other hand beneficially change the intestinal microbial and may be independent of mechanisms of competitive exclusion described earlier.

The potential for control of *E. coli* O157 was introduced by Zhao and co-workers (Zhao et al., 1998). In this study, feeding direct-fed microbials (nonpathogenic *E. coli* and *Proteus mirabilis*) decreased carriage and fecal shedding among those cattle that were artificially inoculated with the pathogen. Specifically, *E. coli* O157 was detected in the control animals for 32 d and in the treated animals for 9 to 17 d. A three-strain combination of nonpathogenic *E. coli* reduced *E. coli* O157 in feces 8 to 30 d after treatment (Tkalcic et al., 2003). Competitive exclusion products containing nonpathogenics require FDA approval prior to use but appear promising in preliminary studies.

Brashears and co-workers (Brashears, Jaroni, and Trimble, 2003) performed benchtop studies and identified specific strains of *Lactobacillus acidophilus* inhibitory to *E. coli* O157. Candidate strains were subjected to a series of tests to determine those best suited to commercial production and survival in the bovine intestinal environment. In a series of small-pen feedlot studies, the specific strain that appears to be most efficacious based on the current body of literature is NP51 (formally known as NPC747). Brashears and colleagues (Brashears, Galyean, Loneragan, Mann, and Killinger-Mann, 2003) evaluated NP51 and another strain, NP35 (formally known as NPC750) in cattle that were housed in five-head pens. On average, *E. coli* O157 was 49 percent less likely to be recovered from steers receiving NP51 compared to controls (OR = 0.51, 95 percent, CI 0.3 to 0.8, $P < 0.01$). The effect of NP35 was not as marked, as only a 30 percent reduction in recovery was detected relative to the controls (OR = 0.7, 95 percent, CI 0.5B1.1, $P = 0.12$). Both *Lactobacillus* strains reduced hide contamination at harvest ($P < 0.05$). In a follow-up study performed in the same feedlot, Younts and colleagues observed a similar benefit with a 58 percent reduction in fecal prevalence in animals that were administered NP51 (Younts-Dahl et al., 2004). Importantly, strain-specific efficacy and strain interaction was evident; when NP51 was co-administered with another strain of *Lactobacillus,* the benefit was not as apparent. This was particularly so if NP51 was administered at a lower dose (10^6 versus 10^9 NP51 cells per head per day). NP51 efficacy was also observed in carriage of *E. coli* O157 on hides.

Lactobacillus dose and strains appear to be important in selecting the appropriate product for efficacious preharvest interventions. Strain selection was emphasized by the earlier work carried out by Brashears and co-workers (Brashears, Jaroni, and Trimble, 2003) in which they evaluated approximately 650 candidate strains. Through a systematic series of studies, only five strains were identified for further evaluation, one of which was NP51. The practical importance of strain variation in observed response is that if a direct-fed microbial strain is selected without any supporting research then conceivably, *E. coli* O157 carriage could inadvertently be increased.

A dose response in NP51 efficacy was evaluated by Younts-Dahl and colleagues (Younts-Dahl et al., 2005). They evaluated three doses, which were 10^7, 10^8, and 10^9 NP51 cells per head. Averaged over exposure time, 10^9 cells per animal per day was associated with the greatest reduction. On average, *E. coli* O157 was 77 percent less likely to be recovered from animals administered 10^9 cells compared to controls (OR = 0.23, 95 percent, CI 0.1 to 0.4, $P < 0.01$). The lower doses also resulted in reduced shedding compared to the controls yet their benefit was not as great when averaged over time. If, however, only the slaughter time-point was considered, all doses demonstrated similar efficacy. As with the previous studies the results of which are outlined earlier, NP51 reduced hide carriage of *E. coli* O157 at harvest. The investigators also evaluated another strain in this study (NP45); animals that were coadministered NP51 and NP45 were more likely to shed *E. coli* O157 in the feces than controls (although not statistically significant; OR = 1.53, 95 percent, CI 0.6 to 4.0, $P = 0.37$).

This product, NP51, is commercially available; company estimates indicate that approximately 10 to 15 percent of the nation's fed cattle are fed this product (personal communication, Douglas Ware, Nutrition Physiology Corporation).

Another potentially promising direct-fed microbial is an *Enterococcus*-based product. As yet, no peer-reviewed studies in which this product has been evaluated are available. However, company data, some of which are derived from small-pen feedlot studies, show promise. In one of the company's reports, animals fed this candidate direct-fed microbial were 56 percent less likely to have *E. coli* O157 recovered from their feces than controls. In challenge studies, they reported a one to two log reduction in *E. coli* O157 in challenged animals compared to controls.

A substantial body of literature now exists about the effect of direct-fed microbials on *E. coli* O157 carriage. Most of this data concerns specific strains of *Lactobacillus acidophilus,* and NP51 has consistently shown significant reductions in fecal shedding and hide carriage. More field-based evaluation of *Enterococcus*- and *E. coli*-based direct-fed microbials is warranted prior to adoption. That said, however, preliminary data are promising and these direct-fed microbials may be targeted interventions appropriate for a systematic multihurdle, farm-to-fork approach to *E. coli* O157 control.

IMMUNOMODULATION

Development of an efficacious vaccine would hold great practical application within the beef industry because (1) cattle are frequently restrained throughout their lives for the specific purpose of vaccine administration; (2) consequently, incorporation of an *E. coli* O157 vaccine into an existing vaccine program would be straightforward; and (3) vaccines have potential for most sectors of the industry whereas some interventions, such as dietary additives, only have application for confined animal-feeding operations. Despite the potential utility of an *E. coli* O157 vaccine to the beef industry, the ability to elicit an effective immune response within the gastrointestinal tract should not be underestimated. For example, natural exposure does not induce protective immunity. Therefore, a vaccine must use specific characteristics of *E. coli* O157 and present the appropriate antigens in a manner that will induce immunity. Even though there are many obstacles to developing an efficacious vaccine, it appears there may be a viable preharvest intervention once regulatory approval is granted.

Investigators administered a vaccine containing type III secreted proteins to feedlot cattle on three occasions. Each administration was 21 days apart (Potter et al., 2004). Averaged over the feeding period subsequent to administration of vaccine for the third time, *E. coli* O157 was recovered from the feces of 8.8 and 21.3 percent of vaccinates and controls, respectively. The authors reported a vaccine efficacy of 58.7 percent, that is, 58.7 percent lower risk of recovery of *E. coli* O157 in vaccinates versus controls. Subsequent studies in which two doses of the vaccine were administered have shown beneficial effects on *E. coli* O157 colonization (data unpublished). The effect of this product on hide carriage was not reported. As shown earlier (Figure 2.3), pen-level carriage of *E. coli* O157 on hides is proportional to pen-level fecal shedding. As such, the effect of vaccination of hide carriage might be implied. A different product has been evaluated in a commercial feedlot study; the investigators of this study found that the vaccine reduced the prevalence of *E. coli* O157 in feces and on hides at harvest (personal communication, Keith Belk, Colorado State University).

Sodium Chlorate

Sodium chlorate is a suicide substrate for certain bacteria within the gastrointestinal tract. It is metabolized within facultatively anaerobic bacteria to highly toxic chlorite (Anderson et al., 2001a). In preliminary studies of cattle and sheep challenged with *E. coli* O157, sodium chlorate effectively reduced pathogen carriage by two to three logs in feces and at various sites in the gastrointestinal tract (Callaway et al., 2002; Callaway et al., 2003a). Authors have demonstrated the bacteriocidal effects of sodium chlorate on *E. coli* O157: H7 and *Salmonella* Typhimurium DT104 both in vitro and in the intestines of experimentally infected pigs (Anderson et al., 2000; Anderson et al., 2001a,b; Anderson et al., 2004), poultry (Byrd et al., 2003); sheep (Callaway et al., 2003a), and cattle (Callaway et al., 2002). Some of these studies used cattle and sheep that were artificially inoculated with specific strains of *E. coli* O157 and the animals were fed potassium nitrate to up-regulate the nitrate reductase enzyme while other studies were performed benchtop. Due to these factors, extrapolation to an expected benefit in naturally infected feedlot cattle not consuming supplemental potassium nitrate is tenuous. However, there is no reason to suspect that sodium chlorate would not be an efficacious preharvest intervention.

A potential benefit beyond *E. coli* O157 control is that sodium chlorate could be used for a variety of foodborne pathogens that are facultatively anaerobic such as *Salmonella* and *Shigella*. This product needs regulatory approval before it can be used. Once approval is attained, it is likely that sodium chlorate will be an effective preharvest intervention.

Neomycin Sulfate

E. coli O157 is a bacterium that, like all other bacteria, is susceptible to certain antimicrobial drugs. It is probable, therefore, that colonization within the gastrointestinal tract might be modified by the use of certain efficacious antimicrobial drugs. Of a large bank of *E. coli*

O157 isolates recovered from naturally infected animals over a series of studies, all were susceptible to neomycin sulfate (personal communication, Robert Elder, Seaboard Farms, Inc.).

Neomycin sulfate has shown to be effective in two commercial feedlot studies. In one study, it was administered in the water for two days; fecal shedding and hide contamination were markedly reduced in harvest-ready cattle ($P < 0.01$ for both models; unpublished data; Figure 2.3). In the control and treated cattle, *E. coli* O157 was recovered from 22.1 and 0.4 percent of feces and 50.0 and 2.5 percent of hides, respectively. This represents a 98.2 and 95.0 percent reduction in fecal and hide recovery, respectively. Hide contamination is generally thought of as representative of cumulative fecal contamination over time and because samples were collected approximately 24 hours after the second day's administration of neomycin sulfate, it is unclear why such a significant reduction was detected on hides. It is uncertain why this was observed; the reasons could be due to (1) the inhibitory effect of neomycin in feces on hides reduced the likelihood of recovery of viable *E. coli* O157 cells; or (2) hide carriage being more dynamic than is generally accepted. In another commercial feedlot study (personal communication, Keith Belk, Colorado State University), a similar benefit of neomycin sulfate on recovery of *E. coli* O157 from feces and hides was observed.

Neomycin sulfate is approved for in-feed or in-water administration to cattle for the control or treatment of colibacillosis, and because of its route of administration, it can be applied to entire groups of animals. However, it is not specifically labeled to reduce *E. coli* O157 reduction. As such, an FDA review to change the label will be required before neomycin sulfate can be used to control *E. coli* O157 in cattle. As the use of antimicrobial drugs, particularly in-feed products, in animal agriculture is becoming increasingly contentious, adoption of neomycin sulfate as a preharvest intervention will likely generate much debate. The conventional applications of antimicrobial drugs in animal agriculture have been to improve animal health or the efficiency of production. The use of neomycin sulfate would be to reduce carriage of *E. coli* O157 in animals destined for human consumption. In other words, it would be administered for a potential human health benefit. The argument concerning neomycin sulfate use, therefore, will have to weigh the potential benefits for public health against the potential costs in terms of favoring development of resistance.

SUMMARY

It is plainly evident that several interventions show considerable efficacy. Other candidate interventions that may be beneficial but are at present in the early developmental phase are beyond the scope of this chapter. Currently, the only available intervention that has been broadly evaluated is the NP51-containing direct-fed microbial. It seems likely that within a short period of time, vaccine technologies to reduce carriage of *E. coli* O157 will be available. These will provide producers with products that best suit their particular production practices.

Other interventions that are presently available, such as the *Enterococcus*-based direct-fed microbials, require additional validation before one can recommend their use. A considerable amount of time and money is required to be invested on those products that need FDA approval.

Despite the challenges associated with discovery, development, and implementation of preharvest interventions, the future is promising for effective on-farm control of *E. coli* O157. Implementation of efficacious preharvest interventions should reduce the in-plant microbial failures. In other words, on-farm control of *E. coli* O157 complements in-plant interventions and increases their efficiency. This should improve the safety of beef, reduce the public exposure to this pathogen, and improve public health. More important, the successful development of preharvest interventions for *E. coli* O157 provides proof of the concept that other pathogens that may arise in the future can be effectively controlled.

LITERATURE CITED

Anderson, R.C. et al. 2000. Bactericidal effect of sodium chlorate on *Escherichia coli* O157:H7 and *Salmonella* typhimurium dt104 in rumen contents in vitro. *J. Food Prot.* 63: 1038-1042.

———. et al. 2001a. Effect of oral sodium chlorate administration on *Escherichia coli* O157:H7 in the gut of experimentally infected pigs. *Int. J. Food Microbiol.* 71: 125-130.

———. et al. 2001b. Effect of sodium chlorate on *Salmonella* Typhimurium concentrations in the weaned pig gut. *J. Food Prot.* 64: 255-258.

———. et al. 2004. Effect of drinking-water administration of experimental chlorate ion preparations on *Salmonella enterica* serovar Typhimurium colonization in weaned and finished pigs. *Vet. Res. Commun.* 28: 179-189.

Angulo, F.J., V.N. Nargund, and T.C. Chiller. 2004. Evidence of an association between use of anti-microbial agents in food animals and anti-microbial resistance among bacteria isolated from humans and the human health consequences of such resistance. *J. Vet. Med. B. Infect. Dis. Vet. Public Health.* 51: 374-379.

Anonymous. 1994. *Escherichia coli* O157:H7 outbreak linked to home-cooked hamburger—California, July 1993. *MMWR* 43: 213-216.

———. 1997. *Escherichia coli* O157:H7 infections associated with eating a nationally distributed commercial brand of frozen ground beef patties and burgers–Colorado, 1997. *MMWR* 46: 777-778.

———. 2001. *Escherichia coli* O157 in United States feedlots. United States Department of Agriculture.

———. 2002a. Multistate outbreak of *Escherichia coli* O157:H7 infections associated with eating ground beef—United States, June-July 2002. *MMWR* 51: 637-639.

———. 2002b. Outbreak of multidrug-resistant *Salmonella* Newport—United States, January-April 2002. *MMWR* 51: 545-548.

Arthur, T.M. et al. 2004. *Escherichia coli* O157 prevalence and enumeration of aerobic bacteria, Enterobacteriaceae, and *Escherichia coli* O157 at various steps in commercial beef processing plants. *J. Food Prot.* 67: 658-665.

Barkocy-Gallagher, G.A. et al. 2003. Seasonal prevalence of shiga toxin-producing *Escherichia coli,* including O157:H7 and non-O157 serotypes, and *Salmonella* in commercial beef processing plants. *J. Food Prot.* 66: 1978-1986.

Bosilevac, J.M. et al. 2004. Prevalence of *Escherichia coli* O157 and levels of aerobic bacteria and Enterobacteriaceae are reduced when hides are washed and treated with cetylpyridinium chloride at a commercial beef processing plant. *J. Food Prot.* 67: 646-650.

Brashears, M.M., M.L. Galyean, G.H. Loneragan, J.E. Mann, and K. Killinger-Mann. 2003. Prevalence of *Escherichia coli* O157:H7 and performance by beef feedlot cattle given *Lactobacillus* direct-fed microbials. *J. Food Prot.* 66: 748-754.

Brashears, M.M., D. Jaroni, and J. Trimble. 2003. Isolation, selection, and characterization of lactic acid bacteria for a competitive exclusion product to reduce shedding of *Escherichia coli* O157:H7 in cattle. *J. Food Prot.* 66: 355-363.

Brashears, M.M., and G.H. Loneragan. 2005. Pre-harvest interventions in beef cattle and challenges associated with implementation. *Assoc. Food Drug Officials* 69: 30-40.

Byrd, J.A. et al. 2003. Effect of experimental chlorate product administration in the drinking water on *Salmonella* Typhimurium contamination of broilers. *Poult. Sci.* 82: 1403-1406.

Callaway, T.R. et al. 2002. Sodium chlorate supplementation reduces *E. coli* O157:H7 populations in cattle. *J. Anim. Sci.* 80: 1683-1689.

———. et al. 2003a. *Escherichia coli* O157:H7 populations in sheep can be reduced by chlorate supplementation. *J. Food Prot.* 66: 194-199.

———. et al. 2003b. Preslaughter intervention strategies to reduce food-borne pathogens in food animals. *J. Anim. Sci.* 81(Suppl 2): E17-E23.

Devasia, R.A. et al. 2005. Antimicrobial use and outcomes in patients with multi-drug-resistant and pansusceptible *Salmonella* Newport infections, 2002-2003. *Microb. Drug Resist.* 11: 371-377.

Elder, R.O., and J.E. Keen. 1999. Effects of pen cleaning and group vs. individual penning on fecal shedding of naturally-acquired enterohemorrhagic *E. coli* (EHEC) O157 in beef feedlot cattle. In: Conference of Research Workers in Animal Diseases, Chicago, IL.

Elder, R.O. et al. 2000. Correlation of enterohemorrhagic *Escherichia coli* O157 prevalence in feces, hides, and carcasses of beef cattle during processing. *Proc. Natl. Acad. Sci. USA*. 97: 2999-3003.

Fey, P.D. et al. 2000. Ceftriaxone-resistant *Salmonella* infection acquired by a child from cattle. *N. Engl. J. Med.* 342: 1242-1249.

Grimm, L.M. et al. 1995. Molecular epidemiology of a fast-food restaurant-associated outbreak of *Escherichia coli* O157:H7 in Washington State. *J. Clin. Microbiol.* 33: 2155-2158.

Gupta, A. et al. 2003. Emergence of multidrug-resistant *Salmonella* enterica serotype Newport infections resistant to expanded-spectrum cephalosporins in the United States. *J. Infect. Dis.* 188: 1707-1716.

Hancock, D.D., T.E. Besser, D.H. Rice, D.E. Herriott, and P.I. Tarr. 1997. A longitudinal study of *Escherichia coli* O157 in fourteen cattle herds. *Epidemiol. Infect.* 118: 193-195.

Hancock, D.D., D.H. Rice, L.A. Thomas, D.A. Dargatz, and T.E. Besser. 1997. Epidemiology of *Escherichia coli* O157 in feedlot cattle. *J. Food Prot.* 60: 462-465.

Helms, M., S. Ethelberg, and K. Molbak. 2005. International *Salmonella* Typhimurium DT104 infections, 1992-2001. *Emerg. Infect. Dis.* 11: 859-867.

Helms, M., P. Vastrup, P. Gerner-Smidt, and K. Molbak. 2002. Excess mortality associated with antimicrobial drug-resistant *Salmonella* Typhimurium. *Infect. Dis.* 8: 490-495.

Hurd, H.S. et al. 2004. Public health consequences of macrolide use in food animals: A deterministic risk assessment. *J. Food Prot.* 67: 980-992.

Hussein, H.S., and L.M. Bollinger. 2005. Prevalence of shiga toxin-producing *Escherichia coli* in beef cattle. *J. Food Prot.* 68: 2224-2241.

Jackson, L.A. et al. 2000. Where's the beef? The role of cross-contamination in 4 chain restaurant-associated outbreaks of *Escherichia coli* O157:H7 in the Pacific Northwest. *Arch. Intern. Med.* 160: 2380-2385.

Kassenborg, H.D. et al. 2004. Farm visits and undercooked hamburgers as major risk factors for sporadic *Escherichia coli* O157:H7 infection: Data from a case-control study in 5 FOODNET sites. *Clin. Infect. Dis.* 38 (Suppl. 3): S271-278.

Laine, E.S. et al. 2005. Outbreak of *Escherichia coli* O157:H7 infections associated with nonintact blade-tenderized frozen steaks sold by door-to-door vendors. *J. Food Prot.* 68: 1198-1202.

LeJeune, J.T., T.E. Besser, and D.D. Hancock. 2001. Cattle water troughs as reservoirs of *Escherichia coli* O157. *Appl. Environ. Microbiol.* 67: 3053-3057.

LeJeune, J.T. et al. 2004. Longitudinal study of fecal shedding of *Escherichia coli* O157:H7 in feedlot cattle: Predominance and persistence of specific clonal types despite massive cattle population turnover. *Appl. Environ. Microbiol.* 70: 377-384.

Loneragan, G.H., and M.M. Brashears. 2005a. Effects of using retention-pond water for dust abatement on performance of feedlot steers and carriage of *Escherichia coli* O157 and *Salmonella* spp. *J. Am. Vet. Med. Assoc.* 226: 1378-1383.

———. 2005b. Pre-harvest interventions for the control of *E. coli* O157 in feedlot cattle. *Meat Sci.* 71: 72-78.

Naylor, S.W. et al. 2003. Lymphoid follicle-dense mucosa at the terminal rectum is the principal site of colonization of enterohemorrhagic *Escherichia coli* O157:H7 in the bovine host. *Infect. Immun.* 71: 1505-1512.

Ostroff, S.M. et al. 1990. A statewide outbreak of *Escherichia coli* O157:H7 infections in Washington State. *Am. J. Epidemiol.* 132: 239-247.

Pearce, M.C. et al. 2006. Prevalence and virulence factors of *Escherichia coli* serogroups o26, o103, o111, and o145 shed by cattle in Scotland. *Appl. Environ. Microbiol.* 72: 653-659.

Phillips, I. et al. 2004. Does the use of antibiotics in food animals pose a risk to human health? A critical review of published data. *J. Antimicrob. Chemother.* 53: 28-52.

Potter, A.A. et al. 2004. Decreased shedding of *Escherichia coli* O157:H7 by cattle following vaccination with type iii secreted proteins. *Vaccine.* 22: 362-369.

Ransom, J.R. et al. 2003. Prevalence of *Escherichia coli* O157 in feedlot cattle feces, on their hides and on carcasses from those cattle. Proceedings of the 10th Meeting of the Int'l. Symp. of Vet. Epid. and Econ., Vina Del Mar, Chile.

Renter, D.G. et al. 2005. Prevalence, risk factors, O serogroups, and virulence profiles of shiga toxin-producing bacteria from cattle production environments. *J. Food Prot.* 68: 1556-1565.

Sargeant, J.M., M.W. Sanderson, D.D. Griffin, and R.A. Smith. 2004. Factors associated with the presence of *Escherichia coli* O157 in feedlot-cattle water and feed in the Midwestern USA. *Prev. Vet. Med.* 66: 207-237.

Sargeant, J.M., M.W. Sanderson, R.A. Smith, and D.D. Griffin. 2004. Associations between management, climate, and *Escherichia coli* O157 in the faeces of feedlot cattle in the Midwestern USA. *Prev. Vet. Med.* 66: 175-206.

Smith, D. et al. 2001. Ecological relationships between the prevalence of cattle shedding *Escherichia coli* O157:H7 and characteristics of the cattle or conditions of the feedlot pen. *J. Food Prot.* 64: 1899-1903.

Smith, D.R. et al. 2005. Use of rope devices to describe and explain the feedlot ecology of *Escherichia coli* O157:H7 by time and place. *Foodborne Pathog. Dis.* 2: 50-60.

Tkalcic, S. et al. 2003. Fecal shedding of enterohemorrhagic *Escherichia coli* in weaned calves following treatment with probiotic *Escherichia coli*. *J. Food Prot.* 66: 1184-1189.

Varma, J.K. et al. 2005. Antimicrobial-resistant nontyphoidal *Salmonella* is associated with excess bloodstream infections and hospitalizations. *J. Infect. Dis.* 191: 554-561.

Younts-Dahl, S.M., M.L. Galyean, G.H. Loneragan, N.A. Elam, and M.M. Brashears. 2004. Dietary supplementation with *Lactobacillus*- and *Propionibacterium*-based direct-fed microbials and prevalence of *Escherichia coli* O157 in beef feedlot cattle and on hides at harvest. *J. Food Prot.* 67: 889-893.

Younts-Dahl, S.M. et al. 2005. Reduction of *Escherichia coli* O157 in finishing beef cattle by various doses of *Lactobacillus acidophilus* in direct-fed microbials. *J. Food Prot.* 68: 6-10.

Zhao, T. et al. 1998. Reduction of carriage of enterohemorrhagic *Escherichia coli* O157:H7 in cattle by inoculation with probiotic bacteria. *J. Clin. Microbiol.* 36: 641-647.

Chapter 3

Beef Safety During Slaughter, Fabrication, and Further Processing

Sally L. Flowers Yoder
Margaret D. Hardin
William R. Henning
Catherine N. Cutter

INTRODUCTION

Meat processors have an ethical responsibility to provide wholesome products to consumers. From the moment the live animal sets foot on the slaughter floor until the beef carcass is fabricated into primals and throughout processing, there are hazards that must be controlled. To fulfill this responsibility, plant employees must be aware that the procedures performed on the slaughter floor and in the fabrication and processing rooms can have a tremendous impact on the safety of the beef products that reach the end user. Currently, every meat plant under federal inspection conducts daily operations under Hazard Analysis and Critical Control Point (HACCP) systems. Each plant develops its own HACCP plan focused on prevention of problems to assure the safety of beef products through control of chemical, physical, and biological hazards.

Large, Small, and Very Small Plants

The U.S. meat and poultry industry comprises approximately 6,000 establishments dedicated to harvest, fabrication, processing,

Handbook of Beef Safety and Quality

doi:10.1300/5640_03

and/or purveying. The USDA classifies plants into one of three categories based on the number of employees and annual revenue. About 5 percent or 300 of these plants are classified as large plants as they employ more than 500 workers, whereas small plants (38 percent, 2,300 establishments) employ between 10 and 500 people. Plants considered very small employ 10 or fewer employces and generate an average annual revenue of $2.5 million or less.

HACCP Implementation

In the interest of public health, it has become unacceptable for contamination with pathogens to persist in muscle foods. The Pathogen Reduction: Hazard Analysis and Critical Control Point Systems final rule of 1996 mandated the implementation of Hazard Analysis and Critical Control Point plans for all meat and poultry establishments, established performance standards for the incidence of *Salmonella* in raw and ground meat products, required periodic testing by the company for generic *E. coli,* and required the development and implementation of Sanitation Standard Operating Procedures (SSOPs) in each plant (USDA-FSIS, 1996c). Employees also should follow the good manufacturing practices (GMPs) that are prescribed within individual meat plants (Katsuyama and Humm, 1995). By following these practices and adhering closely to the plant's HACCP plan, the chemical, physical, and biological hazards associated with beef should be under control at all times. Generally, biological hazards require the most consideration because of the potential for widespread outbreaks of illness in humans.

Primary Processing

Several steps must take place in the slaughter facility to convert a live beef animal to a beef carcass. These steps are described in a general manner, although there is some variation among plants due to existing workspace, available equipment, religious practices, and other factors. During slaughter, exsanguination takes place after stunning and it is the first step in the conversion of muscle to meat. Exsanguination also is the first opportunity during slaughter for microbes to contaminate the carcass, unless captive bolt stunning or some other invasive procedure is used to render the animal unconscious (Aberle et al., 2001). Moreover, microbes can gain access to the bloodstream through the opening made by severing the major blood vessels with a contaminated knife.

Next, the integument (hair, hide, and hooves) is removed. In some plants, chemical dehairing or hide washing may take place prior to exsanguination; then the hide is removed. Hide removal can be achieved by manual skinning or an automated hide puller. When performed manually, the employee should minimize cross-contamination and be careful to keep the skinning knife clean and avoid letting the hide touch the carcass surface as it is being cut away. Automated hide removal can release dust and debris into the ambient air as the hide is pulled away from the carcass by a hydraulic arm. Some of these particles could be desiccated manure that has the potential to carry viable cells of *Escherichia coli* O157:H7, which could be deposited on an otherwise clean carcass surface or processing equipment (Delazari et al., 1998; Elder et al., 2000).

Typically, the head is removed after the carcass has been skinned. At some point, the tail is removed as well. Removal of viscera and pluck from the body cavity, or evisceration, follows. Fecal matter can be released onto carcass surfaces if the intestinal tract is punctured or if the bung is not properly tied. In some plants, carcasses are steam vacuumed for removal of fecal matter and other debris. The carcass is then sawed or split into halves. As a preventive measure for controlling bovine spongiform encephalopathy (BSE), specified risk materials, such as the spinal cord, brain, and skull, must be removed from cattle that are 30 months of age or older (USDA-FSIS, 2004).

At the end of primary processing, the carcass is visually inspected for contamination by employees, undergoes knife trimming as necessary, and is then washed thoroughly with water. In most plants, an intervention step follows the water wash. Interventions may include a sanitizing step with an antimicrobial substance or surface-heat pasteurization. After the final wash, carcasses are placed in a chill cooler or hot box to control the lowering of carcass temperature. Beef carcasses are often cooled by conventional chilling or spray chilling. Conventional chilling, or air chilling, involves heat transfer from the carcass surface to the surrounding air with or without the use of fans under refrigeration conditions. Spray-chilled carcasses are showered with cold water (1°C or higher), which may contain chlorine, in a refrigerated room. After carcasses have been sufficiently chilled, they may undergo quality and/or yield grading prior to fabrication into primals.

Types of Interventions

Currently, there are several types of antimicrobial treatments or interventions shown to substantially reduce bacteria levels on meat and poultry carcasses, and approved for use to meet HACCP requirements. The most common carcass decontamination strategies include water washing, knife trimming, chemical treatments, and moist heat under pressure or vacuum. When considering which interventions to implement, plants of all sizes should consider the expenses of purchase, maintenance, and daily operation (fixed costs); available space and manpower; and documented antimicrobial effectiveness. Some interventions, which are well suited for large plants, are difficult or impossible to implement in small or very small plants.

As part of HACCP, the USDA Food Safety and Inspection Service (FSIS) requires the removal of all visible fecal contamination, ingesta, or milk from beef carcasses, by knife trimming or vacuuming with hot water or steam (USDA-FSIS, 1996b). These carcasses are then subject to re-inspection and must meet finished-product standards applied by plant employees and verified by FSIS inspectors before the carcasses can enter the chiller. Despite the use of sanitary dressing procedures to reduce or eliminate contamination during slaughter and processing, pathogenic bacteria still may be present on beef carcasses (USDA-FSIS, 1996a). Therefore, antimicrobial treatments or interventions are recommended by USDA to reduce the incidence of pathogens in raw beef products.

To reduce spoilage and pathogenic bacteria, large and small beef plants have incorporated antimicrobial treatments or interventions into their slaughter operations that have been shown to greatly reduce the levels of any bacteria that may be present (for a review, see Dickson and Anderson, 1992; Siragusa, 1995; Bolder, 1997; Dorsa, Cutter, and Siragusa, 1997). Employing at least one such treatment will not, by itself, solve the problem of contamination with pathogenic bacteria. However, such treatments are a step among many that can reduce the risk of raw product reaching the consumer with hazardous levels of pathogenic bacteria (USDA-FSIS, 1996a).

CHEMICAL AND PHYSICAL HAZARDS

In general, a food safety hazard can be any agent or property (biological, chemical, or physical) that compromises the safety of a food for human consumption (9 CFR 417.1, NACMCF, 1998). The meat processor and plant employees should be aware of the chemical and physical hazards that are likely to be present in all areas of the meat plant. The HACCP plan must address those hazards that are reasonably likely to occur. While biological hazards are of greatest concern, since they are capable of causing widespread foodborne illness affecting a greater proportion of individuals, chemical and physical hazards also have been associated with foodborne illness and injury. Following is a description of chemical and physical hazards that are likely or known to be of concern in meat-processing facilities. Biological hazards will receive greater emphasis for the remainder of this chapter.

Chemical Hazards

A wide variety of chemical hazards may be present in any meat-processing plant. The chemicals that are most likely to impact beef safety during slaughter, fabrication, and processing originate from incoming cattle, processing aids and ingredients, plant maintenance and plant sanitation (Katsuyama, Jantschke, and Gombas, 1999). As a minimum, the HACCP team should consider all of these sources when performing a hazard analysis.

Beef cattle that enter the plant typically come from reputable feedlots, ranches, or other sources. Livestock producers in the United States are permitted to use hormones and antibiotics in beef production. Growth hormones allow for rapid weight gain while optimizing beef quality (USMEF, 2005a). Antibiotics are approved for use by the Food and Drug Administration for the prevention, treatment, and maintenance of healthy animals (USMEF, 2005b). However, it is the responsibility of the stockman who uses growth promoters, antibiotics, and other chemicals to do so, exactly as indicated by the drug manufacturer, in accordance with the label instructions, and to meet the recommended requirements for withdrawal before sending an animal to slaughter. Moreover, FSIS inspectors routinely monitor food animals for drug residues. In 2001, most of the drug residue violations were traced back to the use of animal drugs at levels that exceed the legal limit or failure to allow sufficient withdrawal time prior to slaughter (USDA-FSIS, 2001). The National Cattlemen's Beef Association, through its Beef Quality Assurance Program, provides guidance and educational materials for producers regarding the judicious use of antimicrobials in cattle to avoid violative residues in food (NCBA, 2001).

In addition to incoming cattle, the plant itself may be the origin of significant chemical hazards. The infrastructure and machinery in meat plants require periodic maintenance to prevent the introduction of these undesirable compounds. Processing equipment, floors, and walls must be cleaned and sanitized as dictated in plant SSOPs. Employees must also wash their hands, personal protective equipment, knives, and other equipment throughout the day. Furthermore, many plants may employ one or more chemical interventions to reduce biological hazards on beef surfaces and in equipment surfaces during processing. For these reasons, forethought is necessary to prevent contact between product and employees and the substances that could be harmful if consumed (e.g., lubricants, paints, cleaners, sanitizers, and excessive concentrations of antimicrobial compounds, food additives, and processing aids; Katsuyama, Jantschke, and Gombas, 1999). The establishment must also ensure that only approved chemicals are used in the facility and that they are used in accordance with label and regulatory recommendations with appropriate written SSOPs. Employee training is essential to prevent misuse and to protect both the employees and the final product.

Incoming nonmeat ingredients, in the form of packaging materials, restricted ingredients, such as nitrates and nitrites, as well as allergens, also may bring chemical hazards into the production facility. While letters of guarantee from suppliers will help to ensure that the chemical safety of these ingredients has been addressed prior to receipt into the facility, it is the responsibility of the establishment to inspect the incoming vehicles and products, provide adequate and appropriate storage, and assure their proper use and labeling.

In recent years, increasing attention has been directed at food allergens. Allergens are proteins that may elicit a specific immune response by the individual that comes in contact with or consumes the food (Katsuyama, Jantschke, and Gombas, 1999). Responses by the susceptible individual upon exposure to an allergen can range from a mild reaction, such as a rash, to more severe reactions such as anaphylaxis, involving the respiratory and circulatory systems, and even death. The most common foods known to cause an allergenic response are often referred to as the "big 8" and have been reported to account for more than 90 percent of the allergenic reactions in adults (Katsuyama, Jantschke, and Gombas, 1999; Deibel et al., 1997). The "big 8" include peanuts, tree nuts (walnuts, pecans, etc.), eggs, dairy (milk, cheese, etc.), soy, wheat, fish, and shellfish (lobster, shrimp, crab, etc.). When processing allergenic and nonallergenic products within the same facility, processors must depend on the prevention of cross-contamination between ingredients, equipment, employees, and product to control this hazard.

The effective control of allergens in a processing facility begins with proper identification of the allergen at receiving and proper storage of allergen-containing ingredients, work-in-process (WIP), and returned product away from allergen-free ingredients (Katsuyama, Jantschke, and Gombas, 1999; Deibel et al., 1997). Scheduling the processing of allergen-containing products after non-allergen-containing products or on a separate production day also will help to limit the potential for cross-contamination. Appropriate labeling of production batch sheets and designating areas where and when an allergen is being used can help prevent cross-contamination. Additional control steps include the attentive cleaning of equipment and proper handling of work apparel (frocks, gloves, sleeves, etc.) after exposure to potential allergens. Employee training is also critical to the control of allergen-related issues. Furthermore, rapid test kits for the detection of allergens in foods are available and could be used to verify process control. Lastly, proper labeling of the final product is essential to properly identify the allergenic ingredient to the consumer.

Physical Hazards

The unintentional introduction of foreign matter into foods also could indicate a physical hazard. Physical hazards associated with beef safety generally are linked to incidents of personal injury, and not public health. Katsuyama and Jantschke (1999) have discussed many specific physical hazards that could occur on the slaughter floor and in the fabrication room. At slaughter, incoming animals may enter the facility with hazards such as injection needles. Furthermore, processing equipment that comes in contact with the carcass is usually made of metal or plastic. For example, teeth from a saw or the tip of a knife can break off, become lodged in the carcass tissue, and later cause injury to a consumer (Katsuyama and Jantschke, 1999). Pieces of plastic may tear or break away from bags, films, or equipment and become attached to or embedded in carcass surfaces only to be discovered post-processing by a consumer. Wood splinters (from wooden-handled equipment, beams, pallets, etc.) and glass shards (from unprotected lighting and glass-covered gauges, thermometers, etc.) also could create physical hazards associated with beef if these materials are present in areas where carcasses or primals are processed.

Physical hazards of concern also include materials such as bone fragments, hair, as well as insects and insect parts (Katsuyama and Jantschke, 1999). Visual inspection of incoming raw materials including meat trim and pieces, packaging materials and spices can help control physical hazards that may put the consumer at risk or damage processing equipment. During processing, physical hazards may also come from meat grinders, dicers, injector needles, and materials used for stuffing and packaging. An assessment of physical risks to beef safety may necessitate the redesign of the process and facility, replacement of plant fixtures, and/or the development or retraining of employees on plant GMPs.

Employees can also be a significant source of foreign objects and personal equipment such as meat hooks, knives, thermometers, pens and pencils, hair, fingernails, and jewelry, which can be hazardous for the final product if employees do not receive proper training or fail to adhere to plant GMPs. Many plants use frocks with inner pockets and snaps instead of buttons to prevent the introduction of foreign objects into edible product. The use of in-line metal detectors or X-ray detection equipment, bone removal systems, and SOPs for visual inspection of equipment (grinder checks, blender checks, needle checks) and product during the day will help to reduce, remove, and control physical hazards. Frequent monitoring of detection equipment, preventive maintenance of facilities and equipment by plant maintenance, and constant training and observation of employees by plant quality assurance and management are necessary to reduce issues associated with physical hazards.

BIOLOGICAL HAZARDS

Individuals who are most susceptible to contracting foodborne disease include the elderly (65 years or older), children who are five years or younger, pregnant women, and individuals with weakened immune systems. The Centers for Disease Control and Prevention estimate that approximately 76 million cases occur each year resulting in 5,000 deaths (CDC, 2003). Moreover, disease incidents from five common foodborne pathogens (*Campylobacter* spp., nontyphoidal *Salmonella, E. coli* O157:H7, *E. coli* non-157:H7 STEC, and *Listeria monocytogenes*) incurred an annual public health cost of almost $6.9 billion in 2000 (USDA-ERS, 2004).

Foodborne illnesses that are traced back to the consumption of fresh meats and poultry products are commonly caused by *E. coli* O157:H7, *Salmonella* spp., or *Campylobacter* spp. The possibility of cross-contamination from raw to ready-to-eat (RTE) products within a facility may present concerns with the previously mentioned pathogens and in particular, *Listeria monocytogenes*. As a result of cross-contamination, *L. monocytogenes* can become a concern in further processed, fully cooked beef products. *Shigella* spp., *Staphylococcus aureus,* Noroviruses, Hepatitis A, and bovine spongiform encephalopathy (BSE) are noteworthy, yet less common biological threats to beef safety.

Escherichia coli *O157:H7*

In the early 1980s *E. coli* O157:H7 was regarded as a rare serotype that was identified initially by the Centers for Disease Control and Prevention (CDC) in a 50-year-old woman who suffered from severe abdominal cramps and bloody diarrhea (Riley et al., 1983). Symptoms of human infection with *E. coli* O157:H7 directly associated with this pathogen may include hemorrhagic colitis, hemolytic uremic syndrome (HUS), thrombotic thrombocytopenic purpura (TTP), or death (Griffin and Tauxe, 1991). A dose of approximately 10 cells has been documented to cause infection (Jay, 2000). *E. coli* O157:H7 illnesses can be life-threatening in children, the elderly, and other individuals with weak immune systems.

The bovine hide and gastrointestinal tract have been well established as significant reservoirs for *E. coli* O157:H7. The virulence of *E. coli* O157:H7 is mediated through shiga or shiga-like toxins, which are also known as verotoxins or verocytotoxins (Wieler, Bauerfeind, and Baljer, 1992). The cells that line bovine intestines lack a receptor that binds the Shigatoxin, so these animals are not susceptible to the organism or toxin. However, the pathogen can attach to and invade human intestinal epithelia, as well as produce Shigatoxin, resulting in cellular death. These events may lead to hemorrhagic colitis or HUS in susceptible populations (Wells et al., 1991). In advanced stages of the illness, the severe symptoms of TTP or HUS can occur and cause death (Griffin and Tauxe, 1991).

Salmonella *spp.*

Salmonella spp. have been implicated in meatborne illnesses for several decades. To cause illness in humans, *Salmonella* cells must be ingested. Common symptoms that are associated with *Salmonella* infection include nausea, vomiting, diarrhea, and severe abdominal pain (Jay, 2000). In some salmonellosis cases, severe dehydration can lead to death of infected individuals. As few as 100 cells of *S.* Eastbourne can cause illness while the average infectious dose of other *Salmonella* strains can be 10^5 to 10^6 CFU/mL (Jay, 2000). At the other extreme, an individual would have to ingest approximately 10^9 to 10^{10} cells of a less virulent strain, such as *S.* Pullorum, before the onset of illness (Jay, 2000).

Of the more than 2,300 serotypes of *Salmonellae, S.* Typhimurium and *S.* Enteriditis are estimated to cause 43.5 percent of the human cases of salmonellosis (CDC, 1998). Of these two species most common to humans, *S.* Typhimurium poses the most critical hazard to beef safety because it is the serotype most frequently isolated (15.7%) from beef (Sarwari et al., 2001). This pathogen has been isolated from feces and rumen contents (Gay, Rice, and Steiger, 1994; McEvoy et al., 2003). *S.* Montevideo, *S.* Anatum, *S.* Kentucky, and *S.* Thompson complete the list of the five most common *Salmonella* serotypes isolated from beef (Sarwari et al., 2001).

Campylobacter *spp.*

Campylobacter spp. are one of the most common causes of foodborne illness. Many cases of campylobacteriosis have been linked to the cross-contamination of prepared foods with raw or undercooked foods and with infected food handlers (Doyle, 2004). Outbreaks of foodborne illness associated with *Campylobacter* spp. are generally associated with drinking raw milk or unchlorinated water, mishandling during preparation of raw poultry, or consumption of undercooked poultry or poultry products (NACMCF, 1998; Altekruse and Swerdlow, 2002).

Disease symptoms are typically less severe than illnesses caused by *E. coli* O157:H7 or *Salmonella* spp. (Shallow et al., 2001). Campylobacteriosis is diagnosed by fecal culture following the onset of severe gastrointestinal upset including abdominal cramps, diarrhea, and vomiting. More severe secondary manifestations of *Campylobacter* infection include the Guillian-Barré syndrome, which can cause acute neuromuscular paralysis, long-term disability, or death in humans, and Reiter's syndrome, which is often characterized by chronic joint pain (Stern, Line, and Chen, 1992).

When compared with other Gram-negative bacteria, *Campylobacter* spp. are considered fragile organisms and, thus, easier to eradicate. *Campylobacter* spp. have been isolated from feces and the intestinal tracts of food animals. When exposed to an environment outside the host animal, the survival and growth of *Campylobacter* spp. can be inhibited by atmospheric oxygen concentrations, competition with other microflora for nutrients, and suboptimal temperatures (>30°C; Doyle, 2004; Stern, Line, and Chen, 1992). Due to the difficulty in isolating these organisms, *Campylobacter* spp. have been identified recently as a major meatborne pathogen (Lammerding et al., 1988). *C. jejuni* and *C. coli* are the two species that are most often linked to illness from the consumption of muscle foods. Even though data are insufficient to support performance standards for *Campylobacter* spp. in the meat and poultry industry, this organism may be the target of future regulation (Lammerding et al., 1988).

Listeria monocytogenes

Listeria monocytogenes is a foodborne pathogen that has been isolated from a broad range of sources, including some 37 species of domestic and wild mammals, and 17 species of birds, soil, and silage (FDA, 1992). Infection with *L. monocytogenes* can lead to listeriosis, which is characterized by septicemia and meningitis in immunocompromised individuals and can cause miscarriage or stillbirths in pregnant women (FDA, 1992).

This pathogen can contaminate carcass surfaces during slaughter, especially if exposed to intestinal contents or feces (Doyle, 1994). *Listeria* is tolerant to desiccation, water activity < 0.93, pH as low as 4.4, refrigeration and freezing temperatures, and salt concentrations as high as 10 percent (Jay, 2000; Lovett and Twedt, 2004). *L. monocytogenes* is an important concern for further processing due to the widespread prevalence of the organism and the ability of the pathogen to survive in a food-processing environment. However, strict sanitation and employee hygiene must be maintained to prevent cross-contamination during slaughter and fabrication. Moreover, meat plants that perform slaughter and further processing within the same facility must take extra precautions to prevent cross-contamination between raw product and raw product processing areas and RTE product and associated processing operations.

Elimination of *Listeria* from the food-processing environment has been found to be extremely difficult as the organism is constantly being introduced by incoming raw materials and employees. A program to control *Listeria* spp. includes thorough and effective cleaning and sanitation of facilities and equipment, separation of raw and RTE, training and strict adherence to GMPs, and monitoring of the environment, equipment, and employees through an environmental sampling program for *Listeria* spp.

Staphylococcus aureus

Staphylococcus aureus is a Gram-positive organism that causes gastrointestinal disease by intoxication. The organism itself is killed by mild heat. Outbreaks of foodborne illness occur after consuming the preformed, heat-stable enterotoxin produced in the food by *S. aureus*. The organism is resistant to high salt (up to 15 percent), has a temperature growth range of 7 to 48°C, pH range of 4 to 10, and may grow at a water activity of 0.85. However, the toxin is generally not produced at temperatures less than 10°C or greater than 48°C and at pH below 4.5 (ICMSF, 1996). The onset of disease symptoms (nausea, vomiting, abdominal cramps, and chills) generally occurs within 30 min to 8 hr after the toxin has been ingested (Newsome, 2004).

Staphylococci are found everywhere in the environment. Outbreaks of foodborne illness are generally associated with protein-rich foods, such as meat salads, and with high salt products, such as ham. While the main reservoirs of this organism are the nasal cavity and skin of humans, animals can also be a source of the organism. Animals can carry staphylococci in nasal passages, on the skin, and it also has been isolated from milk associated with bovine mastitis (Wong and Bergdoll, 2002).

However, outbreaks of foodborne illness associated with meat and meat products are most often the result of contamination of cooked meats during preparation by the equipment or the food handler and not from the organism found in the raw meat (Roberts, 1982; Wong and Bergdoll, 2002). While raw meat is a protein-rich food and may contain staphylococci, the organism found on raw meat is not generally a concern. The pathogen does not compete well with other organisms that are native to raw meat. *S. aureus* is readily destroyed by cooking and does not grow well or produce toxin at lower temperatures (Roberts, 1982; Wong and Bergdoll, 2002). However, staphylococci can be a concern in fermented food production during which food is processed at temperatures that favor the growth of *S. aureus* for a sufficient length of time. Manufacturers of these products must adhere strictly to process parameters for the proper use of starter cultures, and fermentation and maturation times and temperatures (ICMSF, 1996).

Shigella *spp.*

Shigella spp. can cause bacillary dysentery, or shigellosis, in humans. Like many foodborne diseases, shigellosis is characterized by diarrhea, vomiting, abdominal pain, and fever (Flowers, 2004). Like *Salmonella* spp. and *E. coli, Shigella* spp. are classified as Enterobacteriaceae (Jay, 2000). However, *shigellae* are not found in any animal reservoirs besides humans (Jay, 2000). Therefore, poor worker hygiene is the main culprit behind beef carcass contamination with *Shigella* spp. (Genigeorgis, 1987).

Viral Hazards

Among the microbiological hazards associated with beef, the threat posed by viruses often receives little attention. There are several reasons why there are so few data. Viruses require a living host for survival and growth. In addition, as contaminants of foods, viruses tend to be present in small numbers (Jay, 2000). Thus, it is difficult to isolate viruses from foods and culture them in the laboratory (Jay, 2000). Because viruses are very selective of the host species that they infect and the tissues that are colonized, there are few viruses that both infect beef tissue and are transmissible to humans (Cliver, 1980). The majority of viral infections associated with beef consumption are due to contamination by food handlers, who are infected with and/or shedding hepatitis A and/or Noroviruses (Cliver, 1980; Fiore, 2004). To reiterate, this is a common mode of transmission for many pathogenic bacteria such as *Shigella* spp., *Listeria*, and *E. coli* O157:H7. In addition, these infections are not considered life-threatening. Individuals infected with hepatitis A may endure symptoms of gastroenteritis for several weeks and experience jaundice, only to recover with lifelong immunity (Fiore, 2004). Norovirus is the new genus name for the group of viruses also known as Norwalk-like viruses and has been attributed as the cause of nearly 50 percent of foodborne incidents of gastroenteritis (CDC, 2005).

Bovine Spongiform Encephalopathy

Bovine spongiform encephalopathy (BSE) is a subacute transmissible spongiform encephalopathy (TSE) of cattle that was first recognized in the United Kingdom in November 1986 (Wells et al., 1987). TSE diseases cause neurodegenerative disorders in the brain and ultimately result in death. They belong to a group of diseases caused by prions, a protein that induces a conformational change in proteins normally present in animals.

A prion is not a living organism by current definitions (i.e., not viral or bacterial). It has been proposed that when a prion is introduced into a normal cell, it converts the normal cell into a diseased cell

(Horn et al., 2001). Horn et al. (2001) further explain that ingested prions may be absorbed across the gut wall at Peyers patches. These are a part of the mucosal associated lymphoid tissue (MALT). It is thought that the MALT presents microorganisms to the immune system in a contained and ideal fashion, facilitating a protective immune response. Prions could be taken up in the same way. Lymphoid cells then phagocytose the particle and travel to other lymphoid sites such as nodes, the spleen, and tonsils. The prion can replicate at these sites. Many of these sites are innervated and eventually the prion gains access to a nerve and then propagates back up the axon to the spinal cord and eventually the brain (Horn et al., 2001).

BSE is an easily eradicable disease. Since the disease is not contagious, preventing the spread of BSE in the cattle population is simple. In 1997, the United States banned the feeding of meat and bone meal from any ruminant (cattle, sheep, goats, etc.) to other ruminants. Hence, if a cow in the United States is infected with BSE, the disease cannot spread or amplify and the first case becomes the last case in the outbreak. Furthermore, live ruminants and ruminant products from Europe were banned from importation to the United States (USDA-APHIS, 2004a).

BSE is of minimal risk to humans. To prevent the transfer of BSE from cattle to humans, the most important control is to keep the prevalence of BSE in the cattle population to zero or near-zero. Beginning in June 2004, the United States expanded a targeted surveillance program for screening brains of cattle at high risk for developing BSE (USDA-APHIS, 2004b). At the time of this writing, nearly 500,000 high-risk cattle have been tested.

Beyond having a very low prevalence of BSE in the U.S. cattle population, keeping the disease from transferring from infected cattle to humans involves preventing the human consumption of tissue from the central nervous system (CNS). Steaks, roasts, and other muscle cuts do not contain infectious nervous tissue. Nevertheless, some people consume CNS tissue and consider it to be a desirable part of their diet. The other way that humans can consume CNS tissue is by unintentional inclusion of brain or spinal cord in processed meats such as ground beef. Long before the single cow was identified with BSE in the United States, processing methods have been evolving to further decrease the risk of CNS tissue being inadvertently included in any processed meat. New processing methods, such as spinal cord removal, progressively reduce the chance for contamination of ground beef (USDA-FSIS, 2004).

BSE risk to cattle and human health in United States is low. The number of BSE cases in cattle worldwide continues to drop. The number of cases in the U.S. cattle herd will remain near-zero, and exposure of BSE to humans will continue to decrease. Additional protective measures such as mandatory identification and trace-back capability for every animal that enters the food chain should provide further assurance to consumers that American producers will continue to provide high-quality and safe food products.

ANTIMICROBIAL INTERVENTIONS

Water Washing

Spray washing employs the use of a hand-held hose with a spray gun or a specially designed cabinet to deliver water to carcass surfaces for a specified time, at a specified temperature and pressure. For smaller meat processors, hand-held hoses are inexpensive and easy to maneuver. Essentially, the employee directs the water stream at the carcass to wash away gross contamination.

Washing cabinets are made of stainless steel and are equipped with either fixed or rotating nozzles to apply the wash or rinse uniformly over the carcass, as well as within the body cavity. For pork, beef, lamb, or veal processing, these cabinets may extend from floor to ceiling, and generally range from six to fifteen feet long. In some processing facilities, meat carcasses also will undergo spray washing as described earlier after evisceration and immediately before refrigeration. The cost of a spray washing cabinet ranges from $30,000 to $100,000, depending upon the cabinet manufacturer and design specifications.

In a study by Reagan and others, water-washing cabinets in six different beef plants operated under a wide range of conditions (1996). Water-flow rates were as low as 605 L/min and as high as 2,683 L/min while water temperatures ranged from 28°C to 42°C. Also, water was applied to carcass surfaces at different pressures (410 to 2,758 kPA) and for varying lengths of time (18 to 39 s; Reagan et al., 1996). While there is variation among plants in the operating conditions of washing cabinets, carcasses washed in the six plants underwent a 1 $\log_{10}$ reduction in aerobic plate counts. This observation parallels the approximate 1 $\log_{10}$ reduction obtained from water washing in a pilot scale carcass washer under experimental conditions (Cutter and Siragusa, 1994).

Water washing can take place at any stage during slaughter. Large and small plants have adequate manpower to assign workers to washing stations at various phases of slaughter, whereas very small plants may delay water washing until the final wash step. Dickson (1995) explored pre-evisceration washing as a means of reducing bacterial attachment to beef surfaces during the latter stages of slaughter. Dickson (1995) demonstrated that the affinity for bacterial attachment to the carcass surface was lowered by a magnitude of 0.7 log CFU/cm^2 fewer aerobic bacteria and Enterobacteriaceae as compared with control carcasses. Overall, water washing should be employed as an important carcass-cleaning step, which sets the stage for sanitizer effectiveness.

Rinsing with Antimicrobial Compounds

Organic acids (e.g., acetic, lactic, citric), chlorinated compounds, or other generally recognized as safe (GRAS) compounds also can be applied using washing cabinets, similar to the ones used for water washing. A less expensive way for smaller processors to apply antimicrobial solutions to carcasses is with a hand-held spray tank, much like the garden sprayers commonly available at hardware stores or home improvement centers. Overall, these rinses can eliminate microscopic contaminants that remain after the final water-wash step. More important, some antimicrobial rinse solutions produce an inhospitable environment for microbes that remain on carcass surfaces after water washing, and for those that may come in contact with the surface further down the slaughter line. The prolonged contact of the antimicrobial either before evisceration, after evisceration, or during the refrigeration process has been demonstrated to improve both the shelf life and safety of the meat (Barkate et al., 1993; Cutter and Siragusa, 1994; Hardin et al., 1995; Reagan et al., 1996; Davey and Smith, 1989; Dickson, 1995; Dormedy et al., 2000). Various chemical compounds used as carcass sanitizers and their efficacy are discussed later.

Organic Acids

The mechanism of the growth inhibition of common meat spoilage bacteria by organic acids can be attributed to more than the effect of lowering pH alone, as reviewed by Ouattara et al. (1997). At pH levels other than established acid dissociation constants (pKa), organic acids do not fully dissociate in aqueous solutions. The dissolved organic acid does lower pH, forcing bacteria to expend energy to actively export the excess protons. Furthermore, undissociated organic acid molecules are able to enter the cell and then dissociate in the cytoplasm where pH is slightly higher than outside the cell (Ouattara et al., 1997). This shift creates a greater influx of unwanted protons in the cytoplasm and causes the bacterium to work even harder to get rid of them. Eventually, the cell becomes energy-deficient, leading to injury or death (Ouattara et al., 1997).

FSIS permits the use of organic acids as solutions up to 2.5 percent for carcass washing prior to chilling (USDA-FSIS, 2005). Lactic, acetic, and citric acids are the organic acids that are commonly used as carcass rinses, with lactic and acetic acids used most often (Cherrington et al., 1991). When applied as a processing aid, within the approved limits for species and product type, there is no labeling requirement (USDA-FSIS, 2005). The complete inactivation of harmful organisms on beef carcass surfaces cannot be guaranteed, although substantial evidence exists that acid rinsing is an effective means of reducing bacterial populations on carcass surfaces. Hence, acid rinsing can provide reductions as a measure of control, but not elimination, in HACCP plans for slaughter, fabrication, and other fresh meat applications (Dormedy et al., 2000).

As lactic acid is a naturally occurring component in many foods, the addition of food-grade lactic acid is generally recognized as safe by the Food and Drug Administration. It is the most commonly used organic acid for carcass decontamination. It may be used at a concentration up to 5 percent at approximately 55°C as a beef carcass rinse before and after chilling (USDA-FSIS, 2005). However, Van Netten and others (1994) recommend its use at 2 percent, which optimizes antimicrobial efficacy without sacrificing quality, or discoloration of the lean surfaces.

Hardin and others measured the efficacy of water washing followed by organic acid (2 percent lactic acid or 2 percent acetic acid) rinsing for the removal of *E. coli* O157:H7 and *S.* Typhimurium from multiple surfaces of beef carcasses (1995). Rinsing with 2 percent lactic acid (pH = 2.2, 40 psi, 55°C, 200 ml for 11 s, distance from sample = 80 cm) eliminated significantly greater numbers of *E. coli* O157:H7 at more carcass locations than 2 percent acetic acid, with no disparity in the removal of *S.* Typhimurium (Hardin et al., 1995).

Ouattara et al. (1997) investigated the effect of various organic acids (including acetic, lactic, and citric acids) on pure cultures of common meat spoilage bacteria. Acetic and lactic acids were able to inhibit growth of most of the organisms studied at minimum concentrations of 0.1 to 0.75 percent (w/v; Ouattara et al., 1997). Citric acid was considered less effective due to the higher minimum inhibitory concentrations (0.2 to >1.0% w/v) observed to be necessary for the growth of six species of spoilage bacteria. However, when these three organic acids were applied to beef carcass tissue in a pilot scale carcass washer (60 psi, 28°C, 4.2 L/min, distance from sample = 17.8 cm, chain speed = 14 m.min, spray nozzle oscillation speed = 80 cycles/min) at different concentrations (1, 3, or 5%), antimicrobial efficacy against *E. coli* O157:H7 and *Pseudomonas fluorescens* was significantly influenced by acid concentration rather than acid species (Cutter and Siragusa, 1994).

Peroxyacetic Acid

Peroxyacids, such as peroxyacetic acid, also can be used in antimicrobial rinses for carcasses and variety meats. This class of compounds implies that two or more of the following compounds are mixed together before direct application to meats: peroxyacetic acid, octanoic acid, acetic acid, hydrogen peroxide, peroxyoctanoic acid, and 1-hydroxyethylidene-1, 1-diphosphonic acid (also known as HEDP; 21 CFR 173.370). Peroxyacetic acid, by itself, is considered a strong oxidizer that is capable of denaturing microbial proteins and disrupting the bacterial cell membrane (Cords and Dychdala, 1993). When used as a 0.02 percent solution, peroxyacetic acid reduced *E. coli* O157:H7 from beef carcass tissue and beef short plates by 1.4

and 1.0 log CFU/cm^2, whereas water washing produced comparable reductions of 1.2 and 0.3 CFU/cm^2 (Ransom et al., 2003).

Mixtures of peroxyacids that have been optimized for use on beef carcasses are commercially available. The plant employee can dilute the mixture with water to achieve a peroxyacetic acid concentration ≤220 ppm and a hydrogen peroxide concentration of ≤75 ppm. When used at various concentrations (200, 600, and 1000 ppm) and temperatures (45 and 55°C) to eliminate *E. coli* O157:H7 and *S.* Typhimurium from chilled or hot beef carcass surfaces, peroxyacetic acid achieved minor reductions when compared with 2 percent or 4 percent lactic acid rinses (King et al., 2005).

Chlorinated Compounds

Chlorine-based compounds that may be used for microbial decontamination include hypochlorous acid (bleach), chlorine dioxide, and acidified sodium chlorite. The common mode of action of these chlorinated compounds is the ability to oxidize (due to available chlorine) components of the microbial cell membrane, resulting in the disruption of energy generation for survival and growth. Chlorinated compounds are the most commonly used sanitizing agent in the food industry because of cost and ease of use (Cords and Dychdala, 1993).

Currently, acidified sodium chlorite (ASC) is permitted for use as a processing aid to decontaminate beef carcasses, parts, and organs, at a range of 500 to 1200 ppm (21 CFR 173.325). Sodium chlorite must be acidified by a GRAS acid (e.g., citric, acetic, or phosphoric) such that the resulting pH is 2.5 to 2.9 (21 CFR 173.325). Castillo et al. (1999) achieved a 4.5 to 4.6 log reduction of *E. coli* O157:H7 and *S.* Typhimurium by spray washing (mean pH = 2.62, 69 kPa, 22.4 to 24.7°C, 140 ml for 10 s) beef carcass tissue with water, followed by spraying 1200 ppm acidified sodium chlorite activated with citric acid. When activated with phosphoric acid and applied under the same physical conditions, the log reduction in bacterial populations by ASC was 3.8 to 3.9 log cycles, while water washing (manual wash: 69 kPa, 1.5 L for 90 s; followed by automated wash: 5 L applied at 1.72 MPa increasing to 2.76 MPa for 9 s) alone afforded a 2.3 log reduction (Castillo et al., 1999).

Chlorine dioxide is another alternative for sanitizing beef carcass surfaces. Used extensively in poultry chiller water, chlorine dioxide retains more oxidizing capacity than chlorine when organic matter is present (Villareal, Baker, and Regenstein, 1990). Unlike many chlorine compounds currently in use, chlorine dioxide is not converted into chloramines, which are toxic to fish, or trihalomethanes, which are harmful to humans at sufficient levels (Cords and Dychdala, 1993). Chlorine dioxide also maintains antimicrobial effectiveness over a broad pH range. When used as a spray wash (520 kPA, 16°C, 10 s) to decontaminate excised beef carcass surfaces at 0 to 20 ppm, chlorine dioxide was no more effective than water washing (Cutter and Dorsa, 1995). The abundance of organic matter (lean, fat, as well as microbes) on beef carcass surfaces may overpower the oxidizing capacity of chlorine dioxide to inactivate microorganisms as dissociated chlorine loses efficacy once bound to organic matter (Siragusa, 1995). Overall, chlorinated compounds are not as popular as other beef carcass sanitizers (Siragusa, 1995).

Ozonated Water

Ozone has a long history of use as an antimicrobial for drinking water, given that the first drinking water treatment plant to rely on ozone for disinfection was built in the Netherlands in 1893 (Brink, Langlais, and Reckhow, 1991). The gentle bubbling of ozone gas into water generates ozonated water, which can then be used to disinfect foods and food-handling surfaces. Essentially, the mode of action is the release of a third oxygen atom from an unstable ozone molecule. The singular oxygen atoms make contact with organic and inorganic debris and rapidly oxidize them. The remaining oxygen gas remains dissolved in the water or enters the atmosphere. A more thorough review of ozone chemistry is presented by Bablon et al. (1991).

Reagan and associates tested the efficacy of ozonated water, hydrogen peroxide water washing, and knife trimming on experimentally contaminated beef carcasses (1996). Ozone, hydrogen peroxide, and water washing resulted in an approximately 1 log reduction of microbial populations (Reagan et al., 1996). The results of a more recent study agree with those of Reagan and others (1996). Excised carcass surfaces were treated with 95 ppm ozonated water for 30 s or washed with water (28°C) for 9 s with very little difference in effectiveness (Castillo et al., 2003). The effectiveness of ozone as a carcass decontaminant may be stifled by the overwhelming presence of organic material on a carcass surface, since ozone can oxidize components of bacterial cell as favorably as lean or adipose tissue (Castillo et al., 2003).

While ozonated water may not provide the same level of disinfection as other interventions, some plants apply it to food contact surfaces and equipment as a sanitizer after cleaning. For plants that decide to use ozonated water for the dual purposes of plant sanitation and carcass sanitation, ozone can be used on all meat products in accordance with good manufacturing practices (USDA-FSIS, 2005).

Cetylpyridinium Chloride

Cetylpyridinium chloride (CPC) is a quaternary ammonium compound that has recently been approved for use as a poultry carcass rinse and may one day be approved for use on red meat carcass surfaces (USDA-FSIS, 2005). Several authors have briefly reviewed the antimicrobial activity of this compound from a meat-processing perspective (Cutter et al., 2000; Lim and Mustapha, 2004). The chloride and cetylpyridinium ions give it hydrophilic and hydrophobic properties, respectively. This ampiphilic nature may enable the molecule to enter the bacterial cell, damage the membrane, and cause death by leakage of intracellular contents. In the first study to investigate the potential of cetylpyridinium chloride as a beef surface sanitizer, *E. coli* O157:H7 and *S.* Typhimurium were immediately eliminated by 5 to 6 log CFU/cm^2 on lean surfaces and by >2.5 log CFU/cm^2 on adipose surfaces using a 1 percent solution (Cutter et al., 2000).

When used as a sanitizing rinse at 0.5 percent, CPC brought about a 4.8 log CFU/cm^2 and 2.1 log CFU/cm^2 reduction of *E. coli* O157:H7 on beef carcass surfaces and beef short plates, respectively (Ransom et al., 2003). When Ransom et al. (2003) compared this CPC treatment to 2 percent acetic acid, 2 percent lactic acid, 0.02 percent peroxyacetic acid, 0.02 percent acidified sodium chlorite among other germicidal agents, they concluded that CPC was the most effective at eliminating *E. coli* O157:H7 on beef. Other investigators have evaluated the efficacy of CPC as a beef hide decontaminant. Bosilevac et al. (2004) used a pressurized sprayer, held 65 cm from the hide (the entire body except for the head), to deliver 1 percent CPC at 500 psi for 3 minutes to remove as much hide debris as possible and then again for one minute immediately prior to stunning. APC, Enterobacteriaceae, and the prevalence of *E. coli* O157:H7 were reduced by 1.5 log CFU/100 cm^2, 1.1 log CFU/100 cm^2, and 20 percent, respectively (Bosilevac et al., 2004). More research will be needed to support these findings before CPC is approved for direct antimicrobial applications on beef carcasses.

Activated Lactoferrin

Lactoferrin can be isolated from milk, saliva, tears, and white blood cells (Shin et al., 1998). When transformed to a more active form by a patented process, activated lactoferrin can be an efficacious antimicrobial agent (Naidu, 2001; patent 6,172,040). When used as a sanitizing spray, it can be applied electrostatically to beef carcasses as a GRAS solution containing no more than 2 percent activated lactoferrin (USDA-FSIS, 2005). Furthermore, FSIS permits the use of this compound as a processing aid in a multihurdle approach that consists of 1 g activated lactoferrin per dressed beef carcass preceding a water wash and lactic acid rinse. There are strong claims that a 2 percent activated lactoferrin solution can displace pathogens that are tightly bound to meat surfaces and, thus, physically impede recontamination (Naidu et al., 2003). To date, no published data are available to support these claims.

Steam Vacuuming

Another technological intervention used in the meat slaughter process is the steam vacuum sanitizer. The steam vacuum sanitizer may employ a stream of hot water (70-80°C) to moisten and heat treat the carcass surface; a vacuum to remove the moisture along with visible and microscopic contamination from the carcass surface; and steam to sanitize the stainless-steel head. All material vacuumed from the surface is collected into a large container and emptied regularly. The meat industry positions steam vacuum sanitizers throughout the plant to ensure removal of visible contaminants at many steps throughout the slaughter process. Studies have indicated that steam vacuuming can effectively reduce the levels of microbial contamination (Dorsa et al., 1996; Dorsa, Cutter, and Siragusa, 1997; Kochevar et al., 1997; Castillo et al., 1999). A steam vacuum sanitizer costs from $15,000 to $20,000 per station and is used primarily by large or small beef-processing establishments.

Steam Pasteurization

Steam pasteurization employs a multistep process involving water removal, heating, and cooling of meat carcass surfaces, all done within an enclosed cabinet. Similar to the spray-washing cabinets described previously, these enclosed, stainless-steel systems are 14 to 36 feet long and extend from the ceiling to the floor. After evisceration and before refrigeration, carcasses may undergo spray washing to apply hot water (76 to 96°C) or an antimicrobial; pass through high-velocity fans (air velocity = 119 km/h, air volume = 170 m^3/min, drying time = 20 s) to remove residual moisture from the surface; and enter the steam-pasteurization cabinet (Nutsch et al., 1997; Phebus et al., 1997). Within the cabinet, steam (95°C) is applied directly to the carcass surface for several seconds to obtain a carcass surface temperature of approximately 80°C. The heated carcass travels through the cabinet until it is sprayed with cold water to bring the surface temperature down prior to refrigeration. The high temperature is detrimental to pathogenic microorganisms by affecting 4.22 to 4.85 log reductions in *Listeria monocytogenes* Scott A, *E. coli* O157:H7, and *Salmonella* Typhimurium (Phebus et al., 1997). Gill and Bryant (1997) demonstrated a 1.0 log reduction in total aerobic counts and 2.4 and 2.7 log reductions in coliforms and *E. coli,* respectively, on beef carcasses that were steam pasteurized in a commercial plant. Moreover, steam pasteurization does not adversely affect the color or texture of the meat surface (Gill and Bryant, 1997; Phebus et al., 1997). The cost of steam-pasteurization systems ranges from $400,000 for small plants to over a million dollars per unit for large plants.

Chemical Dehairing

Some preliminary studies have elucidated the potential of chemical dehairing as an effective means of pathogen control especially because the hide is known to be a major source of fecal contamination.

A patented chemical dehairing method is summarized by Schnell and others (Clayton and Bowling, 1989; Schnell et al., 1995). The stunned beef animal is sent down the rail to a washing cabinet to be prerinsed with water (8.28 bar, 40°C, 23 sec), sprayed with 10 percent sodium sulfide (3.45 bar, 25°C, 16 sec) followed by a 90 sec dwell before a second spray treatment with 10 percent sodium sulfide (5.52 bar, 25°C, 16 sec) and additional dwell of 60 sec. The carcass is transported further along the cabinet undergoing a water rinse (20.68 bar, 40.5°C, 50 sec) and then a 3 percent hydrogen peroxide rinse (3.45 bar, 17 sec), which neutralizes sodium sulfide. Next, a third water rinse (8.28 bar, 40.5°C, 23 sec) followed by a second rinse with 3 percent hydrogen peroxide (3.45 bar, 17 sec) are applied to the hide surface followed by two more water washes, one at 8.28 bar, 40.5°C, and for 23 s followed by a fresh water rinse at 8.28 bar for 23 s. Upon completion of the dehairing step, the carcass is exsanguinated and subjected to the remainder of the processes associated with primary processing.

Nou et al. (2003) were able to achieve reductions in APC and *Enterobacteriaceae* by 2.0 and 1.8 log (respectively) greater on dehaired carcasses than conventionally processed carcasses. In these experiments, carcasses were dehaired by a proprietary chemical procedure that is similar to the one tested by Schnell and others (1995).

Although dehairing appeared to improve carcass hygiene, this chemical process improved hide hygiene only slightly. Of 240 dehaired carcasses, 67 percent were positive for *E. coli* O157:H7 after dehairing while a significantly higher proportion (88 percent, α = 0.05) of the conventionally processed carcasses were positive for the pathogen (Nou et al., 2003). On the other hand, Schnell et al. (1995) demonstrate that dehaired carcasses had somewhat greater coliform counts (0.3 log) than non-dehaired carcasses, which does not corroborate the advantage of dehairing.

To date, few, if any, plants are employing this technology. It has been speculated that the processing of dehaired carcasses in a plant environment strictly dedicated to dehairing would improve hide hygiene because, theoretically, there would be fewer microbial aerosols in the air and processing equipment would stay cleaner (Schnell et al., 1995; Sofos and Smith, 1998). Additional research is necessary to optimize the processing of chemically dehaired carcasses. There may also be cost barriers to adopting this technology associated with buying equipment, expanding the processing floor, and treating additional wastewater.

Knife Trimming

Knife trimming is frequently performed in plants for the physical removal of fecal matter, ingesta, and milk that visually contaminate carcass surfaces. This regulation is known as zero tolerance. When the antimicrobial effectiveness of knife trimming was compared to water washing and organic acid rinsing, knife trimming was equally as effective as water washing; but, overall it was not as effective as water washing combined with organic acid rinsing (Hardin et al., 1995). In a comparison between knife trimming followed by water washing and water washing alone, no significant differences were detected in the reduction of *E. coli* or total plate count (Gorman et al., 1995). In contrast, Prasai et al. (1995) determined that knife trimming followed by a water wash is more advantageous than knife trimming alone or water washing alone. Washing alone can spread contaminants from one area of the carcass to another, whereas knife trimming alone removes visible debris but fails to eliminate contaminants on untrimmed surfaces (Prasai et al., 1995). In combination, knife trimming removes gross contamination from selected surfaces and thorough water washing makes physical contact with all areas of the carcass to detach and eliminate microbes.

Antimicrobial Interventions Used in Combination

Antimicrobial carcass interventions that consist of two or more surface treatments are generally more effective than water washing or chemical rinsing alone. The efficacy of these multistep approaches can be explained by the hurdle concept (Leistner, 1992). When microorganisms are exposed to a sequence of treatments, the first step inhibits the proliferation of surface microflora by physical removal, cell injury, or cell death. The subsequent treatments can then target the microbes that linger on beef surfaces to further enhance safety (Leistner, 1992).

For example, the efficacy of pre-evisceration washing (345 kPa, 21-54°C, 5.6 s), acetic acid rinsing (2%, 207 kPA, 38-54°C, 5.6 s), warm water washing (2069 kPa, 21-54°C, 20 s), and hot water washing (207 kPa, 80°C, 5.6 s) were performed individually or as sequential spray treatments (Graves Delmore et al., 1998). The same spraying apparatus (9.5 L/min, distance from sample = 25 cm) was used for all treatments. The authors reported a range of log reductions of APC and *E. coli* of 1.2 to 2.2 log CFU/cm^2 and 1.1 to 1.8 log CFU/cm^2, correspondingly, when treatments were performed individually. However, the sequential application of four treatments lowered bacterial populations by 2.8 to 4.3 log CFU/cm^2 (Graves Delmore et al., 1998).

Physical Attributes of Interventions

Pordesimo et al. (2002) provide a compelling argument that engineering variables, or the physical attributes, of intervention treatments should receive more attention in future meat safety research. A review of the literature revealed a lack of standardization of process engineering variables (temperature, pressure, volume, exposure time, distance between food and spray origin, spray pattern and spray nozzle type, number, and orientation) among the numerous studies that have documented the efficacy of water washing and the application of antimicrobial solutions (Pordesimo et al., 2002). Furthermore, there is a need to identify interactions, or the synergistic effects of combining spray wash engineering variables, so that carcass decontamination procedures can optimally harness the kinetic energy of spray droplets (Pordesimo et al., 2002).

The pressure at which a water wash or chemical rinse is applied to a beef carcass surface can influence antimicrobial efficacy. According to Gorman et al. (1995), spray washes applied at pressures greater than 13.79 bar were more effective at removing fecal matter and bacteria on beef adipose surfaces. Studies have indicated the need to demarcate optimal pressure ranges for spray washing since pressure that is too low inadequately removes microbes from meat surfaces, and pressure that is too high can actually drive surface microbes deep into carcass tissues (Pordesimo et al., 2002).

Temperature is a critical factor in the effectiveness of water washing as hot and warm water are generally more lethal treatments than cold water. In fact, washing beef carcasses with cold water was deemed ineffective at reducing aerobic plate counts at various carcass locations and appeared to transfer contaminants from the hindquarter down to the forequarter (Bell, Cutter, and Sumner, 1997). Furthermore, Dorsa, Cutter, and Siragusa (1997) assert that water washes at 70°C or greater are unparalleled at eliminating *E. coli, Salmonella* spp., and background microflora.

Future/Potential Interventions

Current and future research to improve beef safety will continue to explore innovative methods and evaluate novel compounds for carcass decontamination. Recent studies have suggested that irradiation, ultraviolet light, or sodium metasilicate have the potential to develop into carcass interventions for microbiological safety.

In December 1999, the USDA finalized the regulations that govern the irradiation of meat and poultry products (USDA-FSIS, 1999a). According to this final ruling, meat plants establish their own guidelines for irradiation dosage (not to exceed 4.5 kGy for refrigerated product or 7 kGy for frozen product) and the minimum level of pathogen reduction (USDA-FSIS, 1999a). Irradiation has been evaluated for its effectiveness to reduce the prevalence of *E. coli* O157:H7 on beef surfaces. Recently, cutaneous trunci pieces were inoculated with *E. coli* O157:H7 and exposed to E-beam irradiation at 1.0 kGy with 15 mm of penetration (Arthur et al., 2005). The authors report a 2.6 to 2.9 log reduction of *E. coli* O157:H7 when beef surfaces contained a low-level inoculum and a 5.7 to 6.6 log reduction for high-inoculum beef surfaces (Arthur et al., 2005). Additional studies could evaluate the antimicrobial effectiveness of E-beam irradiation on carcass surfaces at other stages during primary processing and fabrication.

In many beef safety experiments, ultraviolet light is used to sanitize excised beef surfaces under laboratory conditions, prior to inoculation and/or antimicrobial treatments (Cutter and Siragusa, 1994; Dorsa et al., 1996; Lim and Mustapha, 2004). Ultraviolet light is classified as radiant energy and it is exposure to radiant energy that accomplishes food irradiation (USDA-FSIS, 1999b). This technology could be a practical means of improving the microbiological safety of beef because carcass contamination is confined to the surface while the internal tissues are considered sterile. Ultraviolet light chambers could be installed along the rail in beef-processing plants to rapidly inactivate surface contaminants from beef carcasses at various stages of slaughter or before fabrication.

Sodium metasilicate has shown promise as an effective means of controlling *E. coli* O157:H7. This compound is GRAS and currently may be used on beef carcasses as a 4 ± 2 percent solution (USDA-FSIS, 2005). Preliminary in vitro data demonstrate complete (>99.99 percent) inhibition of three strains of *E. coli* O157:H7 after 5 to 10 s of exposure to 0.6 percent sodium metasilicate (Weber, O'Brien, and Bender, 2004). Further research is needed to test antimicrobial effectiveness of this alkali on meat surfaces and perhaps on carcasses in a plant setting. Though unclear, the mode of inhibition is thought to be similar to that of trisodium phosphate, another alkaline disinfectant. The combination of high pH and the ability of anions to bind metal ions interfere with bacterial metabolism as well as damage bacterial cell walls and membranes (Weber, O'Brien, and Bender, 2004). Other compounds currently used in the processing and manufacture of other food commodities may cross over to the beef industry for use as antimicrobial agents during slaughter.

FURTHER PROCESSING

Once the carcass has been broken down into primals, retail cuts, and trim, the product may progress to a multitude of further processing steps. Further processing may include but is not limited to chunking, cutting, grinding, mixing, blending, emulsification, tenderizing, injecting and marinating, brining, salting, drying, acidification, and packaging. At any or all of these steps, the addition of ingredients, use of processing equipment, and exposure to processes and employees may influence the overall safety of the final product. It is the responsibility of the production facility to prevent any unsafe or adulterated product from entering commerce where it may adversely affect public health.

Biological, Chemical, and Physical Hazards

The chemical, physical, and biological hazards associated with processing post-fabrication have been addressed in the preceding sections of this chapter. The risk from chemical hazards in further processing are addressed through the proper use of chemicals for maintenance and sanitation, implementation of GMPs that address allergens, and prudent use of additives, such as nitrites, and in process and formulation antimicrobials such as lactates and diacetates. Physical hazards are primarily controlled through preventive maintenance of facilities and equipment as well as GMPs and SOPs to monitor product, equipment, and employees throughout the production process.

The microbiological quality of meat cuts and trim used in further processing is primarily determined by the microbial condition of carcasses and the sanitary conditions under which raw product is processed (Gill, 1998; ICMSF, 1998; NACMCF, 1999). Biological hazards, which may occur after fabrication, are increasingly controlled through GMPs and SSOPs. The physical separation of in-plant manufacturing processes, implementation of antimicrobial interventions, and cold chain management also are critical to the prevention and elimination of microbial contamination.

Besides assessing the microbiological quality of incoming raw materials, specifications for nonmeat ingredients such as water, vegetables, sauces, and spices are also necessary for process control. County or state officials routinely monitor municipal water supplies for the presence of indicator organisms such as generic *E. coli*. These results are sent to the plant and should be reviewed and kept on file. Plants that use well water should have it tested routinely for indicator organisms and chemical contaminants. Incorporation of nonmeat ingredients can be beneficial or detrimental to the microbiological quality of products. While some spices have been associated with foodborne illnesses and product recalls, spices have not been linked to meatborne disease outbreaks. Some herbs and spices, like allspice, clove, cinnamon, and oregano, have intrinsic antimicrobial properties. However, most herbs and spices possess little, if any, antimicrobial activity in food (ICMSF, 1998). In order to minimize negative impacts to the safety of meat products, the use of irradiated herbs and spices is recommended. All nonmeat ingredients used in further processing must be considered in the hazard analysis for their potential to affect the chemical, physical, and biological safety of the product.

Processing Interventions

Extensive research has been conducted related to antimicrobial interventions for carcasses. However, far less research has been conducted and published for post-fabrication processes and products, with most of the substantive research being conducted within the past five years or so. Further complicating the in-plant use and application of post-slaughter interventions is the lack of approvals by USDA and/or FDA for most of the research. Consideration must also be taken to protect the health and safety of employees working near or with antimicrobial compounds.

The presence of *E. coli* O157:H7 in ground beef continues to be a major concern for consumers and regulatory agencies. In an effort to reduce the incidence of illnesses and deaths associated with *E. coli* O157:H7 in ground beef, the USDA Food Safety and Inspection Service has declared this organism an adulterant in fresh ground beef (USDA-FSIS, 1993). Routine testing of trimmings destined for ground beef manufacture and ground beef is mandated. This action has also led to a number of research projects investigating the use of antimicrobials for ground meat and trim. These interventions on beef trim include sprays with hot and cold water, organic acids (lactic, acetic, and gluconic), trisodium citrate, chlorine dioxide, ozonated water and cetyplyridium chloride to reduce varying levels of *E. coli,* aerobic bacteria, coliforms, and *Salmonella* (Dorsa, Cutter, and Siragusa, 1998; Ellebracht et al., 1999; Pohlman et al., 2002; Stivarius et al., 2002a,b; Stivarius, Pohlman, McElyea, and Waldroup, 2002). In other beef research, the direct application of additives to ground beef has been examined. Research with sodium lactate or diacetate, sodium acetate buffered citrate/sodium citrate, and potassium lactate has resulted in limited reductions of *E. coli,* coliforms and aerobic bacteria (Ajjarapu and Shelef, 1999; Egbert et al., 1992; Maca, Miller, and Acuff, 1997). Results of a more recent study examined the effects of lactic acid bacteria on reductions of *E. coli* O157:H7 and *Salmonella* in ground beef. After five days of storage, researchers noted a 5 log reduction in *E. coli* O157:H7 while *Salmonella* was reduced to non-detectable levels (Smith et al., 2005). Carballo et al. (1997) reported up to a 2-log reduction in *E. coli* when ground beef is subjected to high hydrostatic pressure. Kang, Koohmaraie, and Siragusa (2001) demonstrated that multihurdle treatments with water, hot water, hot air, and lactic acid were effective interventions to reduce coliform levels on beef trim.

Likewise, non-intact beef products have also been the subject of antimicrobial interventions studies. By definition, non-intact beef products are manufactured by grinding, injection with solution, mechanical tenderization by needling, cubing, or pounding and restructuring, or reconstructing (NCBA et al., 2005). The translocation of microorganisms from the surface to the interior portions of meat products may pose a risk if pathogens are present on the meat surface initially and are not sufficiently cooked to destroy pathogens inside the product (Hajmeer et al., 2000; Phebus et al., 2000).

An enormous amount of published (and unpublished) data is available to processors of RTE products. This information encompasses a myriad of antimicrobials from lactate/diacetate as a dip, spray, or ingredient, to post-packaging technologies such as hot water post-pasteurization, steam, and radiant heat (Bedie et al., 2001; Gande and Muriana, 2003; Glass et al., 2002; Legan et al., 2004; Muriana et al., 2002). A thorough review of these data should be sought elsewhere.

Nonetheless, the RTE processor must decide whether validation data are appropriate for the process at hand when selecting an antimicrobial strategy. The processor must also ensure that all regulatory approvals have been met before using an antimicrobial. If a processor chooses to use an approved antimicrobial process or product based on published literature, the processor must follow the procedure as outlined in the literature. Also, the processor should confirm that the antimicrobial is properly labeled with usage and safety instructions.

For instance, the application of an antimicrobial solution to the surface of an RTE product necessitates consideration of the proper concentration, which should be maintained if the solution is recirculated or reused throughout the processing day. The processor also should monitor the temperature of the antimicrobial solution and dwell time, if applicable. In the case of spray treatments, process engineering variables should be in order to ensure that the application system operates with the correct nozzles, flow rate, pressure, concentration, and dwell time among other factors.

The most challenging and most effective intervention for control of *Listeria monocytogenes* in RTE beef products includes proper control of the organism in the food-processing environment. Efficacious control measures have been delineated for equipment and facility design, SSOPs and GMPs, and sampling and testing protocols to monitor the effectiveness of these control measures (AMIF, 2003; Cutter and Henning, 2003; ICMSF, 1994; Tompkin, 2002; Tompkin, 2004; Tompkin et al., 1992; Tompkin et al., 1999).

Currently, processing equipment and supplies that have been enhanced with antimicrobial compounds are being developed, ranging from hand soaps and packaging materials to conveyor belts, cutting boards, and floors. Cutter (1999) reported that while antimicrobial-incorporated plastic was effective against various microorganisms, triclosan did not effectively reduce bacterial populations on meat surfaces when stored under refrigerated and vacuum-packaged conditions. Manufacturers of equipment and surfaces impregnated with antimicrobial compounds also need to evaluate the long-term activity of such antimicrobials as the surfaces and equipment are subject to pitting, cracking, and cutting from prolonged exposure to equipment, product, employees, and the processing environment (heat, cold, and humidity).

A number of meat-processing facilities are using interventions in the form of quaternary ammonium and peroxyacid compounds, which are sprayed onto conveyors and/or processing tables during processing. Processors should work with suppliers to maximize the effectiveness of such treatments while balancing cost. Before use, the manufacturers must have prior approval and label instructions that clearly outline specific applications. Treatment of marinades and brines with antimicrobial technologies, such as UV light or ozone, is also being used with varying success.

Furthermore, irradiation has been widely studied and has been reported effective in reducing levels of pathogens in a wide variety of RTE meat products depending on food matrix, irradiation source, and dose. Irradiation is currently approved for use in ground beef and is in use by some ground beef processors. However, availability of the technology, consumer acceptance, and cost remain limitations for many, particularly small, processors.

Retail Operations

While current technologies may reduce the number of pathogens on beef products, the lack of a "kill" step to completely eliminate biological hazards can still pose a risk to consumers (BIFSCo, 2004). As with processors, retailers must depend on temperature control and proper sanitation and personal hygiene (GMPs) to control biological hazards. Temperature control consists of the management of storage facilities and display cases to prevent temperature abuse. Routine inspection of product packaging for leaks, rips, and tears is also recommended. Retailers that grind beef in-store must heed the same precautions that food processors follow. These safety measures include establishing best manufacturing practices, managing a lotting system to keep track of the date and materials used to produce each batch, and having a procedure for handling rework or carryover (BIFSCo, 2004).

SUMMARY

In summary, there are numerous chemical, physical, and biological hazards that pose a threat to beef safety during slaughter and fabrication. The development and implementation of a good HACCP plan provide the backbone for controlling these hazards. Slaughter and fabrication workers should be vigilant while performing processing tasks and be attentive to personal hygiene. Since carcass surfaces are unavoidably contaminated during primary processing, meat plants employ one or more antimicrobial interventions to successfully eliminate harmful microorganisms. The sum of these factors should keep the beef supply safe during the important steps that take place during meat animal slaughter and carcass fabrication.

LITERATURE CITED

Aberle, E.D., J.C. Forrest, D.E. Gerrard, and E.W. Mills. 2001. Conversion of muscle to meat and development of meat quality. Chapter 5 in: *Principles of Meat Science,* 4th ed. Kendall/Hunt Publishing Company, Dubuque, IA, pp. 83-107.

Ajjarapu, S. and L.A. Shelef. 1999. Fate of pGTP-bearing *Escherichia coli* O157: H7 in ground beef at 2 and 10°C and effects of lactate, diacetate, and citrate. *Appl. Environ. Micro.* 65: 5394-5397.

Altekruse, S.F. and D.L. Swerdlow. 2002. *Camplyobacter jejuni* and related organisms. Chapter 6 in: *Foodborne Diseases,* 2nd ed., D.O. Cliver and H.P. Riemann (Eds.). Academic Press, San Diego, CA, pp. 103-112.

AMIF. 2003. Listeria Control Manual: A Comprehensive Step-by-Step Guide for Processors. American Meat Institute Foundation, Supplement. *Food Technol.* December.

Arthur, T.M., T.L. Wheeler, S.D. Shackelford, J.M. Bosilevac, X. Nou, and M. Koohmaraie. 2005. Effects of low-dose, low-penetration electron beam irradiation of chilled beef carcass surface cuts on *Escherichia coli* O157:H7 and meat quality. *J. Food Prot.* 68: 666-672.

Bablon, G., W.D. Bellamy, M. Bourbigot, F.B. Daniel, M. Doré, F. Erb, G. Gordon, B. Langlais, A. Laplanche, B. Legube, G. Martin, W.J. Masschelein, G. Pacey, D.A. Reckhow, and C. Ventresque. 1991. Fundamental aspects. Chapter 2 in: *Ozone in Water Treatment: Application and Engineering,* B. Langlais, D.A. Reckhow, and D.R. Brink (Eds.). American Water Works Association Research Foundation, Denver, CO, and Lewis Publishers, Inc., Chelsea, MI, pp. 11-132.

Barkate, M.L., G.R. Acuff, L.M. Lucia, and D.S. Hale. 1993. Hot water decontamination of beef carcasses for reduction of initial bacterial numbers. *Meat Sci.* 35: 397-401.

Bedie, G.K., K.E. Belk, J.N. Sofos, J.A. Scanga, and G.C. Smith. 2001. Antimicrobials in the formulation to control *Listeria monocytogenes* post-processing contamination on frankfurters stored at 4°C in vacuum packages. *J. Food Prot.* 64: 1949-1955.

Bell, K.Y., C.N. Cutter, and S.S. Sumner. 1997. Reduction of food borne microorganisms on beef carcass tissue using acetic acid, sodium bicarbonate, and hydrogen peroxide spray washes. *Food Microbiol.* 14: 439-448.

BIFSCo (Beef Industry Food Safety Council) and NCBA (National Cattlemen's Beef Association). 2004. Best Practices for Retailer Operations Producing Raw Ground Beef. K.B. Harris (Ed.). International HACCP Alliance, College Station, TX.

Bolder, N.M. 1997. Decontamination of meat and poultry carcasses. *Trends Food Sci. Technol.* 8: 221-227.

Bosilevac, J.M., T.M. Arthur, T.L. Wheeler. S.D. Shackelford, M. Rossman, J.O. Reagan, and M. Koohmaraie. 2004. Prevalence of *Escherichia coli* O157 and levels of aerobic bacteria and *Enterobacteriaceae* are reduced when hides are washed and treated with cetylpyridinium chloride at a commercial beef processing plant. *J. Food Prot.* 67: 646-650.

Brink, D.R., B. Langlais, and D.A. Reckhow. 1991. Introduction. Chapter 1 in: *Ozone in Water Treatment: Application and Engineering,* B. Langlais, D.A. Reckhow, and D.R. Brink (Eds.). American Water Works Association Research Foundation, Denver, CO, and Lewis Publishers, Inc., Chelsea, MI, pp. 1-10.

Carballo, J., P. Fernandez, A.V. Carrascosa, M.T. Solas, and F.J. Colmenero. 1997. Characteristics of low- and high-fat beef patties: Effect of high hydrostatic pressure. *J. Food Prot.* 60(1): 48-53.

Castillo, A., L.M. Lucia, K.J. Goodson, J.W. Savell, and G.R. Acuff. 1999. Decontamination of beef carcass surface tissue by steam vacuuming alone and combined with hot water and lactic acid sprays. *J. Food Prot.* 62: 146-151.

Castillo, A., K.S. McKenzie, L.M. Lucia, and G.R. Acuff. 2003. Ozone treatment for reduction of *Escherichia coli* O157:H7 and *Salmonella* serotype Typhimurium on beef carcass surfaces. *J. Food Prot.* 66: 775-779.

CDC. 1998. 1998 Annual Summary: Table 1. The 20 most frequently reported *Salmonella* serotypes from human sources reported to CDC in 1998 and from nonhuman sources reported to CDC and USDA in 1997. PHLIS Surveillance data. Available at www.cdc.gov/ncidod/dbmd/phlisdata/salmonella.htm, accessed February 16, 2002.

CDC. 2003. Infectious disease information. December 5, 2003. Available at www .cdc.gov/ncidod/diseases/food/index.htm, accessed October 10, 2005.

CDC. 2005. Norovirus: Technical fact sheet. National Center for Infectious Diseases, Respiratory and Enteric Viruses Branch. Available at www.cdc.gov/ ncidod/dvrd/revb/gastro/norovirus-factsheet.htm, accessed August 8, 2005.

Cherrington, C.A., M. Hinton, G.C. Mead, and I. Chopra. 1991. Organic acids: chemistry, antibacterial activity and practical applications. *Adv. Microb. Physiol.* 32: 87-108.

Clayton, R.P. and R.A. Bowling. Inventors. Monfort of Colorado, Inc., assignee. Animal slaughtering chemical treatment and method. US Patent 4,862,557. September 5, 1989.

Cliver, D.O. 1980. Viral hazards in meat. *Proc. Reciprocal Meat Conf.* 33: 63-64.

Cords, B.R. and G.R. Dychdala.1993. Sanitizers: Halogens, surface-active agents, and peroxides. Chapter 14 in: *Antimicrobials in Foods*, 2nd ed., P.M. Davidson and A.L. Branen (Eds.). Marcel Dekker, Inc., New York, pp. 469-537.

Cutter, C.N. 1999. The effectiveness of triclosan-incorporated plastic against bacteria on beef surfaces. *J. Food Prot.* 62: 474-479.

Cutter, C.N. and W.J. Dorsa. 1995. Chlorine dioxide spray washes for reducing fecal contamination on beef. *J. Food Prot.* 58: 1294-1296.

Cutter, C.N., W.J. Dorsa, A. Handie, S. Rodriguez-Morales, X. Zhou, P.J. Breen, and C.M. Compadre. 2000. Antimicrobial activity of cetylpyridinium chloride washes against pathogenic bacteria on beef surfaces. *J. Food Prot.* 63: 593-600.

Cutter, C.N. and W.R. Henning. 2003. Control of *Listeria monocytogenes* in small meat and poultry establishments. The Pennsylvania State University, State College, PA.

Cutter, C.N. and G.R. Siragusa. 1994. Efficacy of organic acids against *Escherichia coli* O157:H7 attached to beef carcass tissue using a pilot scale model carcass washer. *J. Food Prot.* 57: 97-103.

Davey, K.R., and M.G. Smith. 1989. A laboratory evaluation of a novel hot water cabinet for the decontamination of sides of beef. *Int. J. Food Sci.* 24: 305-316.

Deibel, K.E., T. Trautman, T. DeBoom, W.H. Sveum, G. Dunaif, V.N. Scott, and D.T. Bernard. 1997. A comprehensive approach to reducing the risk of allergens in foods. *J. Food Prot.* 60: 436-441.

Delazari, I., S.T. Iaria, H. Riemann, D.O. Cliver, and N. Jothikumar. 1998. Removal of *Escherichia coli* O157:H7 from surface tissues of beef carcasses inoculated with wet and dry manure. *J. Food Prot.* 61: 1265-1268.

Dickson, J.S. 1995. Susceptibility of preevisceration washed beef carcasses to contamination by *Escherichia coli* O157:H7 and *salmonellae*. *J. Food Prot.* 58: 1065-1068.

Dickson, J.S. and M.E. Anderson. 1992. Microbiological decontamination of food animal carcasses by washing and sanitizing systems: A review. *J. Food Prot.* 55: 133-140.

Dormedy, E.S., M.M. Brashears, C.N. Cutter, and D.E. Burson. 2000. Validation of acid washes as critical control points in hazard analysis and critical control point systems. *J. Food Prot.* 63: 1676-1680.

Dorsa, W.J., C.N. Cutter, and G.R. Siragusa. 1997. Effects of steam-vacuuming and hot water spray wash on the microflora of refrigerated beef carcass surface tissue inoculated with *Escherichia coli* O157:H7, *Listeria innocua*, and *Clostridium sporogenes*. *J. Food Prot.* 60: 619-624.

Dorsa, W.J., C.N. Cutter, and G.R. Siragusa. 1998. Bacterial profile of ground beef made from carcass tissue experimentally contaminated with pathogenic and spoilage bacteria before being washed with hot water, alkaline solution, or organic acid and then stored at 4 or 12°C. *J. Food Prot.* 61: 1109-1118.

Dorsa, W.J., C.N. Cutter, G.R. Siragusa, and M. Koohmaraie. 1996. Microbial decontamination of beef and sheep carcasses by steam, hot water spray washes, and a steam-vacuum sanitizer. *J. Food Prot.* 59: 127-135.

Doyle, M.P. 1994. The emergence of new agents of food borne disease in the 1980s. *Food Res. Internat.* 27: 219-226.

Doyle, M.P. 2004. *Campylobacter jejuni* and other species. In: *Bacteria Associated with Food Borne Diseases*. IFT Scientific Status Summary, Available at members.ift.org/NR/rdonlyres/3DEA7A91-DF48–42CE-B195–106B01C14 E 273/0/bacteria.pdf, accessed August 17, 2005.

Egbert, W.R., D.L. Huffman, D.D. Bradford, and W.R. Jones. 1992. Properties of low-fat ground beef containing potassium lactate during aerobic refrigerated storage. *J. Food Sci.* 5: 1094-1037.

Elder, R.O., J.E. Keen, G.R. Siragusa, G.A. Barkocy-Gallagher, M. Koohmaraie, and W.W. Laegreid. 2000. Correlation of enterohemorrhagic *Escherichia coli* O157 prevalence in feces, hides, and carcasses of beef cattle during processing. *Proc. NAS* 97: 2999-3003.

Ellebracht, E.A., A. Castillo, L.M. Lucia, R.K. Miller, and G.R. Acuff. 1999. Reduction of pathogens using hot water and lactic acid on beef trimmings. *J. Food Prot.* 64: 1094-1099.

FDA. 1992. *Foodborne Pathogenic Microorganisms and Natural Toxins Handbook* (Bad Bug Book). Center for Food Safety & Applied Nutrition. Available at www.cfsan.fda.gov/~mow/chap6.html, accessed August 8, 2005.

Fiore, A.E. 2004. Hepatitis A transmitted by food. *Clin. Infect. Dis.* 38: 705-715.

Flowers, R.S. 2004. Shigella. In: *Bacteria Associated with Food Borne Diseases*. IFT Scientific Status Summary. Available at: members.ift.org/NR/rdonlyres/3DEA7A91-DF48–42CE-B195–106B01C14E273/0/bacteria.pdf, accessed August 17, 2005.

Gande, N. and P. Muriana. 2003. Prepackage surface pasteurization of ready-to-eat meats with radiant heat oven for reduction of *Listeria monocytogenes*. *J. Food Prot.* 66: 1623-1630.

Gay, J.M., D.H. Rice, and J.H. Steiger. 1994. Prevalence of fecal *Salmonella* shedding by cull dairy cattle marketed in Washington state. *J. Food Prot.* 57: 195-197.

Genigeorgis, C. 1987. The risk of transmission of zoonotic and human diseases by meat and meat products. In: *Elimination of Pathogenic Organisms from Meat and Poultry*, F.J.M. Smulders (Ed.). Elsevier Scientific Publishers, New York, pp. 111-147.

Gill, C.O. 1998. Microbiological contamination of meat during slaughter and butchering of cattle, sheep and pigs. Chapter 4 in: *The Microbiology of Meat and Poultry*, A. Davies and R. Board (Eds.). Blackie Academic & Professional, London, pp. 118-157.

Gill, C.O. and J. Bryant. 1997. Decontamination of carcasses by vacuum-hot water cleaning and steam pasteurizing during routine operations at a beef packing plant. *Meat Sci.* 47: 267-276.

Glass, K.A., D.A. Granberg, A.L. Smith, A.M. McNamaara, M. Hardin, J. Mattias, K. Ladwig, and E.A. Johnson. 2002. Inhibition of *Listeria monocytogenes* by sodium diacetate and sodium lactate on wieners and cooked bratwurst. *J. Food Prot.* 65: 116-123.

Gorman, B.M., J.B. Morgan, J.N. Sofos, and G.C. Smith. 1995. Microbiological and visual effects of trimming and/or spray washing for removal of fecal material from beef. *J. Food Prot.* 58: 984-989.

Graves Delmore, L.R., J.N. Sofos, G.R. Schmidt, and G.C. Smith. 1998. Decontamination of inoculated beef with sequential spraying treatments. *J. Food Sci.* 63: 890-893.

Griffin, P.M. and R.V. Tauxe. 1991. Epidemiology of infections caused by *Escherichia coli* O157:H7, other enterhemorrhagic *E. coli*, and the associated hemolytic uremic syndrome. *Epidemiol. Rev.* 13: 60-98.

Hajmeer, M.N., E. Ceylan, J.L. Marsden, and R.K. Phebus. 2000. Translocation of natural microflora from muscle surface to interior by blade tenderization. Cattlemen's Day 2000. Kansas State University, Manhattan, KS. Report on Progress 850, pp. 125-126.

Hardin, M.D., G.R. Acuff, L.M. Lucia, J.S. Oman, and J.W. Savell. 1995. Comparison of methods for decontamination from beef carcass surfaces. *J. Food Prot.* 58: 368-374.

Horn, G.H., M. Bobrow, M. Bruce, M. Godert, A. McLean, and J. Webster. 2001. Review of the origin of BSE. London: Department for Environment, Food and Rural Affairs (DEFRA), pp. 1-69.

ICMSF. 1994. Choice of sampling plan and criteria for *Listeria monocytogenes*. International Commission on Microbiological Specifications for Foods. *Int. J. Food Microbiol.* 22: 89-96.

ICMSF. 1996. Microorganisms in foods. Volume 5. *Microbiological Specifications of Food Pathogens.* Blackie Academic & Professional, London.

ICMSF. 1998. Microorganisms in foods. Volume 6. *Microbial Ecology of Food Commodities.* Blackie Academic & Professional, London.

Jay, J.M. 2000. *Modern Food Microbiology,* 6th ed. Aspen Publishers, Inc., Gaithersburg, MD.

Kang, D., M. Koohmaraie, and G.R. Siragusa. 2001. Application of multiple antimicrobial interventions for microbial decontamination of commercial beef trim. *J. Food Prot.* 64: 168-171.

Katsuyama, A.M. and B.J. Humm. 1995. The regulations of HACCP to CGMPs and Sanitation. Chapter 13 in: *Establishing Hazard Analysis and Critical Control Point Programs: A Workshop Manual,* 2nd ed., K.E. Stevenson and D.T. Bernard (Eds.). The Food Processors Institute, Washington, DC, pp. 13-17.

Katsuyama, A.M. and M. Jantschke. 1999. Physical hazards and controls. Chapter 7 in: *Establishing Hazard Analysis and Critical Control Point Programs: A Workshop Manual,* 3rd ed., K.E. Stevenson and D.T. Bernard (Eds.). The Food Processors Institute, Washington, DC, pp. 63-66.

Katsuyama, A.M., M. Jantschke, and D.E. Gombas. 1999. Chemical hazards and controls. Chapter 6 in: *Establishing Hazard Analysis and Critical Control Point Programs: A Workshop Manual,* 3rd ed., K.E. Stevenson and D.T. Bernard (Eds.). The Food Processors Institute, Washington, DC, pp. 53-61.

King, D.A., L.M. Lucia, A. Castillo, G.R. Acuff, K.B. Harris, and J.W. Savell. 2005. Evaluation of peroxyacetic acid as a post-chilling intervention for control of *Escherichia coli* O157:H7 and *Salmonella* Typhimurium on beef carcass surfaces. *Meat Sci.* 69: 401-407.

Kochevar, S.L., J.N. Sofos, R.R. Bolin, J.O. Reagan, and G.C. Smith. 1997. Steam vacuuming as a pre-evisceration intervention to decontaminate beef carcasses. *J. Food Prot.* 60: 107-113.

Lammerding, A.M., M.M. Garcia, E.D. Mann, Y. Robinson, W.J. Dorward, R.B. Truscott, and F. Tittiger. 1988. Prevalence of *Salmonella* and thermophilic *Campylobacter* in fresh pork, beef, veal and poultry in Canada. *J. Food Prot.* 51: 47-52.

Legan, J.D., D.L. Seaman, A.L. Milkowski, and M.H. Vandeven. 2004. Modeling the growth boundary of *Listeria monocytogenes* in ready-to-eat cooked meat products as a function of the product salt, moisture, potassium lactate, and sodium diacetate concentrations. *J. Food Prot.* 67: 2195-2204.

Leistner, L. 1992. Food preservation by combined methods. *Food Res. Intern.* 25: 151-158.

Lim, K. and A. Mustapha. 2004. Effects of cetylpyridinium chloride, acidified sodium chlorite, and potassium sorbate on populations of *Escherichia coli* O157:H7, *Listeria monocytogenes,* and *Staphylococcus aureus* on fresh beef. *J. Food Prot.* 67: 310-315.

Lovett, J. and R.M. Twedt. 2004. *Listeria monocytogenes* and other *Listeria* species. In: *Bacteria Associated with Food Borne Diseases.* IFT Scientific Status Summary. Available at members.ift.org/NR/rdonlyres/3DEA7A91-DF48–42CE-B195–106B01C14E273/0/bacteria.pdf, accessed August 17, 2005.

Maca, J.V., R.K. Miller, and G.R. Acuff. 1997. Microbiological and chemical characteristics of vacuum-packaged ground beef patties treated with salts of organic acids. *J. Food Sci.* 62: 591-596.

McEvoy, J.M., A.M. Doherty, J.J. Sheridan, I.S. Blair, and D.A. McDowell. 2003. The prevalence of *Salmonella* spp. in bovine faecal, rumen, and carcass samples at a commercial abattoir. *J. Appl. Microbiol.* 94: 693-700.

Muriana, P.M., W. Quimby, C.A. Davidson, and J. Grooms. 2002. Postpackage pasteurization of ready-to-eat deli meats by submersion heating for reduction of *Listeria monocytogenes*. *J. Food Prot.* 65: 963-969.

NACMCF, 1998. Hazard analysis and critical control point principles and application guidelines. National Advisory Committee on Microbiological Criteria for Foods. *J. Food Prot.* 61: 762-775.

NACMCF. 1999. FSIS microbiological hazard identification guide for meat and poultry components of products produced by very small plants. August 26, 1999. Available at www.fsis.usda.gov/OA/haccp/kidguide.htm, accessed September 15, 2005.

Naidu, A.S., inventor and assignee, Immobilized lactoferrin antimicrobial agents and the use thereof. US Patent 6,172,040. January 9, 2001.

Naidu, A.S., J. Tulpinski, K. Gustilo, R. Nimmagudda, and J.B. Morgan. 2003. Activated lactoferrin Part 2: natural antimicrobial for food safety. *Agro Food Ind. Hi Tech.* May/June 27-31.

NCBA. 2001. Beef Quality Assurance Program. National Cattlemen's Beef Association. Available at www.bqa.org, accessed September 15, 2005.

NCBA, AMI, NMA, and SMA. 2005. Best practices for pathogen control during tenderizing/enhancing of whole muscle cuts. American Meat Institute, National Meat Association, and Southwest Meat Association. Reviewed by: Beef Industry Food Safety Council, Denver, CO.

Newsome, R.L. 2004. *Staphylococcus aureus*. In: *Bacteria Associated with Food Borne Diseases*. IFT Scientific Status Summary. Available at members.ift.org/NR/rdonlyres/3DEA7A91-DF48-42CE-B195-06B01C14E273/0/bacteria.pdf, accessed August 17, 2005.

Nou, X., M. Rivera-Betancourt, J.M. Bosilevac, T.L. Wheeler, S.D. Shackelford, B.L. Gwartney, J.O. Reagan, and M. Koohmaraie. 2003. Effect of chemical dehairing on the prevalence of *Escherichia coli* O157:H7 and the levels of aerobic bacteria and *Enterobacteriaceae* on carcasses in a commercial beef processing plant. *J. Food Prot.* 66: 2005-2009.

Nutsch, A.L., R.K. Phebus, M.J. Riemann, D.E. Schafer, J.E. Boyer Jr., R.C. Wilson, J.D. Leising, and C.L. Kastner. 1997. Evaluation of a steam pasteurization process in a commercial beef processing facility. *J. Food Prot.* 60: 485-492.

Ouattara, B., R.E. Simard, R.A. Holley, G.J.-P. Piette, and A. Bégin. 1997. Inhibitory effect of organic acids upon meat spoilage bacteria. *J. Food Prot.* 60: 246-253.

Phebus, R.K., A.L. Nutsch, D.E. Schafer, R.C. Wilson, M.J. Riemann, J.D. Leising, C.L. Kastner, J.R. Wolf, and R.K. Prasai. 1997. Comparison of steam pasteurization and other methods for reduction of pathogens on freshly slaughtered beef surfaces. *J. Food Prot.* 60: 476-484.

Phebus, R.K., H. Thippareddi, H. Sporing, J.L. Marsden, and C.L. Kastner. *E. coli* O157:H7 risk assessment for blade-tenderized beef steaks. Cattlemen's Day 2000. Kansas State University, Manhattan, KS. Report on Progress 850, pp. 117-118.

Pohlman, F.W., M.R. Stirvarius, K.S. McElyea, and A.L. Waldroup. 2002. Reduction of *E. coli, Salmonella typhimurium,* coliforms, aerobic bacteria, and improvement of ground beef color using trisodium phosphate or cetylpyridinium chloride before grinding. *Meat Sci.* 60: 349-356.

Pordesimo, L.O., E.G. Wilkerson, A.R. Womac, and C.N. Cutter. 2002. Process engineering variables in the spray washing of meat and produce. *J. Food Prot.* 65: 222-237.

Prasai, R.K., R.K. Phebus, C.M. Garcia Zepeda, C.L. Kastner, A.E. Boyle, and D.Y.C. Fung. 1995. Effectiveness of trimming and/or washing on microbiological quality of beef carcasses. *J. Food Prot.* 58: 1114-1117.

Ransom, J.R., K.E. Belk, J.N. Sofos, J.D. Stopforth, J.A. Scanga, and G.C. Smith. 2003. Comparison of intervention technologies for reducing *Escherichia coli* O157:H7 on beef cuts and trimmings. *Food Prot. Trends.* 23: 24-34.

Reagan, J.O., G.R. Acuff, D.R. Buege, M.J. Buyck, J.S. Dickson, C.L. Kastner, J.L. Marsden, J.B. Morgan, R. Nickelson, II, G.C. Smith, and J.N. Sofos. 1996. Trimming and washing of beef carcasses as a method of improving the microbiological quality of meat. *J. Food Prot.* 59: 751-756.

Riley, L.W., R.S. Remis, S.D. Helgerson, H.B. McGee, J.G. Wells, B.R. Davis, R.J. Hebert, E.S. Olcott, L.M. Johnson, N.T. Hargrett, P.A. Blake, and M.L. Cohen. 1983. Hemorrhagic colitis associated with a rare *Escherichia coli* serotype. *New Eng. J. Med.* 308: 681-685.

Roberts, D. 1982. Bacteria of public health significance. Chapter 9 in: *Meat Microbiology,* W.H. Brown (Ed.). Applied Science Publishers LTD, Essex, England, pp. 319-386.

Sarwari, A.R., L.S. Magder, P. Levine, A. McNamara, S. Knower, G.L. Armstrong, R. Etzel, J. Hollingsworth, and J.G. Morris Jr. 2001. Serotype distribution of *Salmonella* isolates from food animals after slaughter differs from that of isolates found in humans. *J. Infect. Dis.* 183: 1295-1299.

Schnell, T.D., J.N. Sofos, V.G. Littlefield, J.B. Morgan, B.M. Gorman, R.P. Clayton, and G.C. Smith. 1995. Effects of postexsanguination dehairing on the microbial load and visual cleanliness of beef carcasses. *J. Food Prot.* 58: 1297-1302.

Shallow, S., M. Samuel, A. McNees, G. Rothrock, D. Vugia, T. Fiorentino, R. Marcus, S. Hurd, P. Mshar, Q. Phan, M. Cartter, J. Hadler, M. Farley, W. Baughman, S. Segler, S. Lance-Parker, W. MacKenzie, K. McCombs, P. Blake, J.G. Morris, M. Hawkins, J. Roche, K. Smith, J. Besser, E. Swanson, S. Stenzel, C. Medus, K. Moore, S. Zansky, J. Hibbs, D. Morse, P. Smith, M. Cassidy, T. McGivern, B. Shiferaw, P. Cieslak, M. Kohn, T. Jones, A. Craig, and W. Moore. 2001. Preliminary FoodNet Data on the Incidence of Foodborne Illnesses—Selected Sites, United States, 2000. *Morb. Mortal. Week.* 50: 241-246.

Shin, K., K. Yamauchi, S. Teraguchi, H. Hayasawa, M. Tomita, Y. Otsuka, and S. Yamazaki. 1998. Antibacterial activity of bovine lactoferrin and its peptides against enterohaemorrhagic *Escherichia coli* O157:H7. *Lett. Appl. Microbiol.* 26: 407-411.

Siragusa, G.R. 1995. The effectiveness of carcass decontamination systems for controlling the presence of pathogens on the surfaces of meat animal carcasses. *J. Food Safety* 15: 229-238.

Smith, L., J.E. Mann, K. Harris, M.F. Miller, and M.M. Brashears. 2005. Reduction of *Escherichia coli* O157:H7 and *Salmonella* in ground beef using lactic acid bacteria and the impact on sensory properties. *J. Food Prot.* 68: 1587-1592.

Sofos, J.N. and G.C. Smith. 1998. Nonacid meat decontamination technologies: Model studies and commercial applications. *Int. J. Food Microbiol.* 44: 171-188.

Stern, N.J., J.E. Line, and H. Chen. 1992. *Campylobacter.* Chapter 31 in: *Compendium of Methods for the Microbiological Examination of Foods,* 3rd ed. C. Vanderzant and D.F. Splittstoesser (Eds.). American Public Health Association, Washington, DC, pp. 301-310.

Stivarius, M.R., F.W. Pohlman, K.S. McElyea, and J.K. Apple. 2002a. Microbial, instrumental color and sensory color and odor characteristics of ground beef produced from beef trimmings treated with ozone or chlorine dioxide. *Meat Sci.* 60: 299-305.

Stivarius, M.R., F.W. Pohlman, K.S. McElyea, and J.K. Apple. 2002b. The effects of acetic acid, gluconic acid and trisodium citrate treatment of beef trimmings on microbial, color and odor characteristics of ground beef through simulated retail display. *Meat Sci.* 60: 245-252.

Stivarius, M.R., F.W. Pohlman, K.S. McElyea, and A.L. Waldroup. 2002. Effects of hot water and lactic acid treatment of beef trimmings prior to grinding on microbial, instrumental color and sensory properties of ground beef during display. *Meat Sci.* 60: 327-334.

Tompkin, R.B. 2002. Control of *Listeria monocytogenes* in the food-processing environment. *J. Food Prot.* 65: 709-725.

Tompkin, R.B. 2004. Environmental sampling—A tool to verify the effectiveness of preventive hygiene measures. *Mitt. Lebensm. Hyg.* 95: 45-51.

Tompkin, R.B., L.N. Christiansen, A.B. Shaparis, R.L. Baker, and J.M. Shroeder. 1992. Control of *Listeria monocytogenes* in processed meats. *Food Australia* 44: 370-376.

Tompkin, R.B., V.N. Scott, D.T. Bernard, W.H. Sveum, and K.S. Gombas. 1999. Guidelines to prevent post-processing contamination from *Listeria monocytogenes. Dairy Food and Environ. San.* 19: 441-562.

USDA-APHIS. 2004a. Bovine spongiform encephalopathy (BSE). United States Department of Agriculture Animal and Plant Health Inspection Service. Available at: www.aphis.usda.gov/lpa/issues/bse/bse-overview.html, accessed September 27, 2005.

USDA-APHIS. 2004b. USDA BSE surveillance plan: Background on assumptions and statistical inferences. Available at www.aphis.usda.gov/lpa/issues/bse/BSEOIG.pdf, accessed September 27, 2005.

USDA-ERS. 2004. Economics of foodborne disease: Feature. Economic Research Service. Available at www.ers.usda.gov/briefing/FoodborneDisease/features.htm, accessed October 10, 2005.

USDA-FSIS. 1993. Immediate actions: Cattle clean meat program. FSIS correlation packet, interim guidelines for inspectors. Food Safety and Inspection Service, Washington, DC.

USDA-FSIS. 1996a. FSIS Directive 6350.1: Use of knife trimming and vacuuming of beef carcasses with hot water or steam; other beef carcass intervention systems.

USDA-FSIS. 1996b. Notice of policy change; achieving the zero tolerance performance standard for beef carcasses by knife trimming and vacuuming with hot water or steam; use of acceptable carcass interventions for reducing carcass contamination without prior agency approval. *Federal Register* 61, no. 66 (April 4, 1996).

USDA-FSIS. 1996c. Pathogen reduction act; hazard analysis and critical control point (HACCP) systems; final rule. *Federal Register* 61, no. 144 (July 25, 1996).

USDA-FSIS. 1999a. Irradiation of meat food products; final rule. Federal Register 64, no. 246 (December 23, 1999).

USDA-FSIS. 1999b. USDA issues final rule on meat and poultry irradiation. Food Safety and Inspection Service. Available at www.fsis.usda.gov/OA/background/irrad_final.htm, accessed August 17, 2005.

USDA-FSIS. 2001. 2001 National Residue Program Data-Red Book. Available at www.fsis.usda.gov/Science/2001_National_Residue_Data_Red_Book/index.asp, accessed August 17, 2005.

USDA-FSIS. 2004. Prohibition of the use of specified risk materials for human food and requirements for the disposition of non-ambulatory disabled cattle; meat produced by advanced meat/bone separation machinery and meat recovery (AMR) systems; prohibition of the use of certain stunning devices used to immobilize cattle during slaughter; bovine spongiform encephalopathy surveillance program; interim final rules and notice. *Federal Register* 69, no. 7 (January 12, 2004).

USDA-FSIS. 2005. FSIS Directive 7120.1, Amendment 4: Safe and suitable ingredients used in the production of meat and poultry products.

USMEF. 2005a. Backgrounder Antibiotics. Available at www.usmef.org/TradeLibrary/assets/12233/antfactsheetfile/Factsheet%20-%20Antibiotics.pdf, accessed September 15, 2005.

USMEF. 2005b. Backgrounder Hormone Use. United States Meat Export Federation. Available at www.usmef.org/TradeLibrary/assets/12233/horbackgrounderfile/backgrounder%20%20Hormones.pdf, accessed September 15, 2005.

Van Netten, P., J.H.J. Huis in 't Veld, and D.A.A. Mossel. 1994. The immediate bactericidal effect of lactic acid on meat-borne pathogens. *J. Appl. Bacteriol.* 77: 490-496.

Villareal, M.E., R.C. Baker, and J.M. Regenstein. 1990. The incidence of *Salmonella* on poultry carcasses following the use of slow release chlorine dioxide (Alcide). *J. Food Prot.* 53: 465-467.

Weber, G.H., J.K. O'Brien, and F.G. Bender. 2004. Control of *Escherichia coli* O157:H7 with sodium metasilicate. *J. Food Prot.* 67: 1501-1506.

Wells, G.A.H., A.C. Scott, C.T. Johnson, R.F. Gunning, R.D. Hancock, M. Jeffrey, M. Dawson, and R. Bradley. 1987. A novel progressive spongiform encephalopathy in cattle. *Vet. Rec.* 121: 419-420.

Wells, J.G., L.D. Shipman, K.D. Greene, E.G. Sowers, J.H. Green, D.N. Cameron, F.P. Downes, M.L. Martin, P.M. Griffin, S.M. Ostroff, M.E. Potter, R.V. Tauxe, and I.K. Wachsmuth. 1991. Isolation of *Escherichia coli* serotype O157:H7 and other shiga-like-toxin-producing *E. coli* from dairy cattle. *J. Clin. Microbiol.* 29: 985-989.

Wieler, L.H., R. Bauerfeind, and G. Baljer. 1992. Characterization of shiga-like toxin producing *Escherichia coli* (SLTEC) isolated from calves with and without diarrhoea. *Zbl. Bakt.* 276: 243-253.

Wong, A.C.L. and M.S. Bergdoll. 2002. Staphylococcal food poisoning. Chapter 16 in: *Foodborne diseases,* 2nd ed. D.O. Cliver and H.P. Riemann (Eds.). Academic Press, San Diego, CA, pp. 231-248.

PART II: BEEF QUALITY

Chapter 4

The Quality Revolution

Thomas G. Field
Deborah L. VanOverbeke

I'm afraid as great as computers are, they cannot tell you about the quality of your product. The profitability, yes, but not the quality. The human eye, the human experience, is the one thing that can make quality better—or poorer.

Stanley Marcus, former chairman, Neiman-Marcus

The heart of quality is not technique. It is a commitment by management to its people and product—stretching over a period of decades and lived with persistence and passion—that is unknown in most organizations today.

Tom Peters and Nancy Austin,
A Passion for Excellence: The Leadership Difference

Our product is more than beef. It's the smell of sage after a summer thunderstorm, the cool shade of a ponderosa pine forest. It's 80-year-old weathered hands saddling a horse in the Blue Mountains, the future of a 6 year old in a one room school in the High Desert. It's a trout in a beaver built pond; haystacks on an Aspen framed meadow. It's the hardy quail running to join the cattle for a meal, the welcome ring of a dinner bell at dusk.

Oregon Country Beef

Handbook of Beef Safety and Quality

doi:10.1300/5640_04

INTRODUCTION

Perhaps revolution is too strong a descriptor for the quality focus of the U.S. beef industry. In comparison to the Japanese electronics industry, Disney, or 3M, the beef industry has yet to initiate the intense effort required to attain dramatic improvements in the overall quality of its products. As Table 4.1 illustrates, the industry has a long way to go toward achieving the kind of standards of performance that would suggest anything other than "ours is no worse than yours."

There are, of course, examples of individual enterprises, divisions within companies, teams, and supply chain partnerships where a deeply held commitment to quality has indeed had a substantial impact. However, quality cannot be forced on an industry, a business, or an individual. Quality can only be created one day, one product, one customer, and one interaction at a time. Quality is not a goal to be obtained and then checked off on the proverbial list of things to do.

TABLE 4.1. Comparative results of the National Beef Quality Audits: 1991, 1995, 2000, and 2005.

Characteristic	Goal	1991	1995	2000	2005
No brands	100.0	55.5	47.7	49.3	62.0
No horns	100.0	68.9	67.8	77.3	77.7
No manure	100.0	93.2	61.6	18.0	25.8
No bruises	100.0	NA	51.6	53.3	64.8
Liver condemnations	0.0	19.2	22.2	30.3	24.7
USDA Prime and Choice	65.0	55	48	51	57.3
A-maturity carcasses	100.0	93	95	97	97.1
B-maturity carcasses	0.0	7.0	4.9	3.4	1.7
Prime	6.0	2.2	1.3	2.0	2.9
Choice	59.0	52.7	46.7	49.1	54.2
Select	35.0	36.9	46.7	42.3	36.7
Standard	0.0	7.6	4.6	5.6	
Fat thickness (in)		0.59	0.47	0.49	
Carcass weight (lbs)		759	748	787	
Ribeye area (in^2)		12.9	12.8	13.1	
Yield grade		3.2	2.8	3.0	

Source: National Cattlemen's Beef Association (1991, 1995, 2000) and Smith et al. (2006).

Quality results from a deliberate choice of a philosophy about doing business that requires us to battle daily against the forces of cynicism, slipshod products, and workmanship, the easy path of settling for the mundane, and the slow downward spiral created when human beings allow rules, bureaucracy, and the vicelike grip of the status quo to separate our people and our organizations from creativity, uniqueness, and inspiration.

The problem is that quality cannot be defined, much less created via technique, recipe, or formula. Quality is not clinical, statistical, nor quantifiable. Quality is experienced deep in our gut and when people talk about quality products or services they rarely, if ever, use quantifiable terminology. Instead they vividly describe the feel, emotion, passion, or essence of the experience. Quality is measured in smiles, laughter, a quickened heartbeat, and the heightened senses of taste, smell, sight, touch, and sound. Quality is born from the heart rather than from the mind. If the emotions of love, passion, commitment, inspiration, and courage are not interwoven with our vision of quality then it is unlikely, if not impossible, to create anything of lasting value.

Quality is a 24-hours-a-day, 7-days-a-week, 52-weeks-a-year commitment. Quality does not get to take a day off. Quality is a quest for perfection and an investment in persistent improvement. Quality is not a slogan, it is not a management edict, and it is very difficult to achieve and even more difficult to sustain. A commitment to quality requires that we return to the core principle of pride of ownership. And pride of ownership has to radiate from everybody on our team and our customer and supplier's teams, every day, and in everything we do. Quality is not for the fainthearted, and if the last three paragraphs have not caused you to squirm then you have not yet grasped the enormous task of undertaking a quality revolution.

What does it take to pull off a quality revolution? In *Thriving on Chaos*, Tom Peters (1987) outlined the attributes of a quality revolution:

1. *Management is obsessed with quality.* Notice that Peters used the word "obsessed" instead of "interested" or "focused." Obsessed implies a degree of heroic preoccupation with getting it right and suggests that to do anything less is simply unaccept-

able. Quality is a term that rolls off the tongues of most beef industry leaders at every level but until it is the central focus of our efforts and philosophy, its meaning is significantly lessened. Obsession also provides the steel required to withstand the setbacks and to consistently push the organization into a pride of ownership philosophy.

2. *There is a guiding system in place to facilitate the attainment of quality goals.* "Most quality programs fail for one of two reasons—they have system without passion, or passion without system. You must have both," according to Peters (Peters, 1987, p. 101).
3. *Quality is measured.* The process to effectively measure quality involves documenting the cost of poor quality in the entire chain. Results are reported to everyone (cowboys to clerks) in the organization, and measurement is conducted directly by the people who have responsibility for what is being evaluated (and then encouraged to use the data).
4. *Quality is rewarded.* Consequences typically generate action. A note of caution: The reward should be appropriate and structured to enhance dedication to attainment of quality objectives.
5. *Everyone is trained to assess quality.* Techniques such as statistical process control and hazard analysis critical control point are continuously taught at every level of the business. It simply will not do to train management-level employees and to then expect everyone else to deliver.
6. *Multifunction teams are formed to solve problems.* These teams utilize a variety of talents and job skills not only from people within the organization but potentially from suppliers and customers as well to respond to quality issues as they arise. Once the problem is solved, the team is disbanded. Such an approach keeps fresh ideas and the best mix of perspective and talent at the table relative to a specific problem.
7. *Small improvements are celebrated.* It is rare to find a single breakthrough approach that accelerates an organization to new heights. Long-term success is built around a culture of continuous incremental improvement.
8. *Constant stimulation is provided to help people through the mundane times.* Improvement can be painfully slow and it is in-

cumbent upon leadership to help provide the environment necessary to sustain the resolve of team members.

9. *Quality cannot be obtained without the buy-in of suppliers, distributors, and customers.* Perhaps the most important benefit of such an effort is improved communication and the development of trust-based relationships.
10. *As quality rises, costs decline.* Cost-reduction efforts undertaken in a vacuum can seriously disrupt the quality of products or services. Deming (1994) and others have shown the substantial cost savings that occur from systems designed to produce quality as opposed to those that depend on inspection to assure desired levels of quality. Peters (1987) suggests that the primary advantage comes from simplification of protocols, processes, and structure.
11. *Quality improvement is never-ending.* In a free and competitive market, quality is never static—it is either getting better or worse relative to the competitors.

OVERVIEW OF TOTAL QUALITY MANAGEMENT

Total quality management (TQM) is a means by which an industry can meet the needs of its customers. W. Edwards Deming, the father of TQM, outlined the chain reaction resulting from putting TQM into place. The reaction, as outlined by Walton (1986) includes the following steps, which, if put into place, will help to create a quality-driven system:

1. Improve quality.
2. Costs decrease because of less rework, fewer mistakes, fewer delays, snags, better use of machine time, and materials.
3. Productivity improves.
4. Capture the market with better quality and lower price.
5. Stay in business.
6. Provide jobs and more jobs.

TQM is not a quick fix; rather, it is a long-term business concept in which returns for doing things right may not be direct or immediate.

Deming describes the process of TQM as being one where things are done right the first time, but this is done by achieving 14 key principles. Deming's 14 points as described by Walton (1986) are as follows:

1. *Create constancy of purpose for improvement and service.* Change your role from "make money" to "stay in business and provide jobs through innovation, research, constant improvement, and maintenance" (p. 34).
2. *Adopt the new philosophy.* Mistakes and negativism are unacceptable.
3. *Cease dependence on mass inspection.* Quality comes from improvement of a process, not from mass inspection—do away with paying employees to make defects and then to correct them.
4. *End the practice of awarding business on price tag alone.* Seek the best-quality supplier instead of the lowest-priced vendor; work to find a sole supplier for a given item with a long-term relationship.
5. *Improve constantly and forever the system of production and service.* Improvement is not a one-time process; continually seek ways to improve the process, reduce waste, and improve quality.
6. *Institute training.* Instruct people on how to do their jobs; do not let the untrained teach the new employee.
7. *Institute leadership.* Leadership is helping people do a better job and helping them learn; leadership is not telling people what to do.
8. *Drive out fear.* The economic loss from fear is unimaginable. Make people feel secure—comfortable in asking questions and taking a position.
9. *Break down barriers between staff areas.* Ensure that all employees, regardless of department, are working together toward a common goal versus competing among one another.
10. *Eliminate slogans, exhortations, and targets for the workforce.* Since these rarely help people perform to an expected level, let people create their own motivational stimuli.
11. *Eliminate numerical quotas.* Deming states that quotas often represent inefficiency and high cost, only taking numbers into account instead of focusing on quality.

12. *Remove barriers to pride of workmanship.* People are eager to do a good job; too often other people or the quality and reliability of equipment or materials stand in the way. Find a way to remove those barriers.
13. *Institute a vigorous program of education and retraining.* Everyone will have to be educated in the new methods, including teamwork, and logic to production of a quality product; do not assume that they will "hear" about the new system and know what their responsibilities are.
14. *Take action to accomplish the transformation.* This transformation will not happen overnight and will be the result of a dedicated group of people, from every level in the company. The critical mass must be in supporting and understanding the goal of the changes being implemented.

The concept of TQM in the beef industry is implemented via programs such as the Beef Quality Assurance (BQA) Program during the production phases of production and the Hazard Analysis Critical Control Points (HACCP) system during processing. A quality system, such as BQA or HACCP, is a network of interdependent components integrated together to accomplish the mission of the system (Deming, 1994). As such, a quality beef product is not the result of individual effort but the combined efforts and decisions of the packer, retailer, restaurant, and each individual producer.

THE ROLE OF PROFESSIONAL STANDARDS IN INTERNATIONAL TRADE

The events surrounding outbreaks of Foot and Mouth Disease (FMD), Bovine Spongiform Enchephalopathy (BSE), and Avian Influenza (Bird Flu) point to the tremendous challenges of marketing food on a global scale. The disruptions to trade created by both real and perceived risk associated with animal disease issues cannot be ignored.

If nothing else, the U.S. beef industry learned in 2004-2005 that the proverbial horse had left the barn in regard to the globalization of beef trade. The closure of international markets to U.S. beef and the sealing of the Canadian border to live cattle trade in late 2003 and early 2004 cost the U.S. beef industry between $165 and $190 per head (Cattle-Fax, 2005). Much of the loss was incurred as the result of lost markets to items such as short ribs, tongues, and outside skirts that are valued more highly in the international market than in the domestic arena.

The internationalization of food marketing has made it incumbent on policymakers and food systems participants to develop functional standards in the monitoring, analysis, and application of protocols designed to ensure food safety. While harmonization of these standards requires significant effort, consumer confidence and the development of profitable international markets are dependent on just such an effort. Perhaps most importantly, development of international standards that are founded on agreed-upon scientific merit provide the basis for well-defined policy decisions that balance national interests, profitability and sustainability of agricultural production systems, and consumer welfare.

The international standard upon which quality-management systems, process controls, hazard analyses, and verified supply chains are built is the ISO 9000. ISO 9000 provides the common language for quality management and is designed to enhance the ability of management to make better decisions, to better deal with risk, and to provide buyers a higher degree of assurance of quality and safety. The benefits resulting from standardization include the following:

- Reduction of trade disruptions based on process or analytical methodologies;
- Enhance the ability to verify quality, process, and source;
- Recovery of uncaptured efficiencies (Hurburgh and Lawrence, 2002).

The generally accepted international code for the food industry is the Codex Alimentarius. Founded in 1961, the Codex Alimentarius is a body of standards, best practices, and other recommendations (Table 4.2). The organization of the document is described next.

Codex Standards

- Relate to product characteristics such as maximum residue limits;

- Commodity standards include information such as scope, product descriptions, product composition, food additives, contaminants, hygiene, labeling, sampling technique, and analytical methodology.

Codex Codes of Practice

- Define production, processing, manufacturing, transportation, and storage practices for individual or groups of food;
- The goal is to ensure food safety for consumers.

Codex Guidelines

- Two classifications—principles that define policy plus guidelines for the interpretation of the policy.

In principle, the harmonization of food-related laws and regulations should reduce trade barriers, improve agricultural profitability, and enhance food availability to consumers. Uniform codes also enhance the ability of governments to assure that their own citizens are protected from foodborne disease and that their national livestock herds and plant populations are protected from disease.

TABLE 4.2. Summary of the text of the Codex Alimentarius.

Category	**Number of guidelines**
Commodity standards	202
Commodity-related guidelines and codes of practice	38
Food labeling standards and guidelines	7
Food hygiene codes and guidelines	5
Food safety risk assessment guidelines	5
Guidelines on food contaminants	14
Sampling, analysis, inspection, and certification standards	22
Allowable limits for pesticide residues	2,579 (213 pesticides)
Guidelines for food additives	683 (222 food additives)
Limits for veterinary drugs in foods	377 (44 drugs)

Source: Codex Alimentarius, http://www.codexalimentarius.net/web/index_en.jsp.

DEFINING THE ROLE OF QUALITY IN ENHANCING BEEF DEMAND

Domestic Markets

In the United States, USDA quality grades are used to define quality of beef products and are used as a marketing tool for beef products. In 2005, the National Beef Quality Audit determined that in the United States 3 percent Prime carcasses, 19.3 percent upper 2/3 Choice, 35.0 percent lower 1/3 Choice, 37.3 percent Select, 4.1 percent Standard, and 1.3 percent hardboned carcasses were produced (Smith et al., 2006). However, during the same audit, representatives from hotel, restaurant, and retail trade markets indicated that they preferred a distribution of 6 percent Prime, 27 percent upper 2/3 Choice, 32 percent lower 1/3 Choice, and 35 percent Select; with no Standard or hardboned carcasses in the mix (Smith et al., 2000).

The NBQA—2000 also defined the leading quality challenges in the beef industry (Smith et al., 2000). These challenges included the following:

- Low overall uniformity and consistency in cattle, carcasses, and cuts;
- Inappropriate carcass size and weight;
- Inadequate tenderness of beef;
- Insufficient marbling;
- Reduced-quality grade and tenderness due to implants;
- Excess external fat cover;
- Inappropriate USDA quality grade mix;
- Too much hide damage due to brands;
- Too frequent and severe bruises;
- Too frequent liver condemnations.

The industry continues to audit these traits in fed steers and heifers to determine what improvement has been made and assists producers by providing them with strategies and tactics each audit year to improve on the quality of cattle in the industry.

In today's industry, multiple factors are being recognized as quality-demand drivers. These characteristics include process-verified programs, organic programs, natural programs, source-verified programs, and the like. Deming's definition of quality is the *totality of characteristics that comprise value*. To some consumers in the domestic, and even international, markets, organic, natural, source verification, and other niche programs comprise additional value.

While markets depend on quality, it is also important to understand that there will be product variation, no matter what level of quality exists (Deming, 1994). Fortunately the diversity of the marketplace provides the opportunity to effectively market the variability. However, widespread variability in a product niche will ultimately erode demand.

International Markets

The 1994 International Beef Quality Audit pointed to both the strengths and weaknesses of U.S.-produced beef in regard to meeting the needs of international consumers. The primary strengths as perceived by international customers included the following:

1. Superior product availability as the result of the industry's ability to sell specific cuts/items in volume as opposed to whole carcasses only;
2. Ninety-seven percent of those surveyed viewed the eating quality of U.S. beef as more than satisfactory;
3. U.S. beef was viewed as providing a high level of value;
4. Positive image of the U.S. quality grading system;
5. Strong levels of confidence in the level of product safety associated with U.S. beef.

The weaknesses uncovered by the survey included the following:

1. Too much external fat relative to the specifications of the purchase contract;
2. Boxed beef weights as well as the weight of whole-muscle cuts were considered too heavy;
3. Customer service was lacking;
4. U.S. beef cuts contained too much seam fat;
5. Shelf life of U.S. beef was not competitive with other providers;
6. Excessive purge in vacuum-packaged beef as well as problems with package leakage;

7. Poor box/container quality;
8. Poor overall workmanship compared to alternate beef suppliers;
9. Labeling problems, such as labels being difficult to interpret and not being written in the language of the customer's culture.

The authors of the International Beef Quality Audit (1994) summarized their findings with the following five key points:

1. The U.S. industry lacked a commitment to servicing the international marketplace;
2. The quality gap between U.S. beef and its competitors was narrowing;
3. Quality defects were costing the U.S. industry sales;
4. Many of the quality defects were interrelated and thus resulted in system-wide deficiencies in product integrity;
5. The U.S. beef supply chain lacked the coordination, cooperation, and communication to fully meet the demands of the export market.

In the period from 1995 to 2003, the U.S. beef industry transformed its focus on export markets and made significant strides in correcting the shortcomings addressed in the 1994 audit. By the end of 2003, the equivalent of approximately 10 percent of the total U.S. beef supply was being sold via export markets.

The challenges brought by animal health issues such as BSE have resulted in serious setbacks to the ability of the U.S. beef industry to develop its export market share. The Japanese market recently opened to U.S. beef as of September 2006; the current trade agreement and emerging trade agreements that will go into effect upon trade normalization with other countries will create a new set of rules. North American producers will have to verify that their cattle are younger than 21 months of age at the time of harvest under the regulations of the Beef Export Verification (BEV) Program to qualify their cattle for trade with Japan. Under the proposed rules, processors will have to participate in a Quality System Assessment (QSA) to accomplish the following objectives:

1. Ensure that cattle purchased or received from outside vendors and used in the program meet the age restrictions.

2. Select suppliers based on their ability to conform to the standards.
3. Develop and implement a system of inspection to make sure that the product conforms to the age requirement.
4. Utilize a documented protocol to select, evaluate, and reevaluate suppliers.
5. Maintain records of supplier evaluations and any corrective action taken.
6. Maintain records to provide evidence of conformity to the receiving process. (Lawrence, 2005)

The job of recapturing lost market share in the international market is not an easy one. However, most certainly, the ability of the U.S. industry to regain its role as a major beef exporter will hinge on the ability to provide consumer confidence through a transparent system of tracing back and source verification. However, it is important to recognize that these will be the requirements for reentry and will not guarantee a return to the dominant position held by the industry prior to 2004. Addressing the very real concerns such as those cited by foreign buyers in the 1994 International Beef Quality Audit will hold the key for international market recovery for the U.S. beef industry.

MOVING THE INDUSTRY FROM COMMODITY TO VALUE-ADDED

Agriculture has a long-established tradition of commodity orientation fueled by the cyclic behavior of supply and demand in a marketplace where product differentiation is lacking and pricing is established on a lot average basis. The problem with commodity marketing is that by definition the average producer under such a system will only break even over time—hardly the desired outcome of an entrepreneurial business. In essence, commodities trade under the general theme that "one producer's product is not different than that of any other producer" (Field, personal communication). As such, under this system a producer becomes a price taker.

The value-added approach to food production recognizes that not only do individual farms produce distinctive raw ingredients but also there is the unique differentiation of consumers into specific market niches.

DEVELOPING QUALITY-FOCUSED SUPPLY CHAINS

Quality-focused supply chains can be as simple as working together as cow-calf producers to produce similar calves for market; feedlots working together to supply a branded beef program; packers working to sort carcasses into specific quality groups for direct marketing to restaurants, retail markets, or end users; or further processors obtaining a product meeting specifications that they use to develop a specialty product. In the United States, the United States Department of Agriculture's Agricultural Marketing Service (USDA, AMS) has provided several means by which producers and processors can develop and verify quality-focused supply chains.

The USDA, AMS monitors two types of programs: beef carcass certification programs and USDA Process Verified programs. Carcass certification programs provide certification of beef carcasses for a variety of carcass specifications for marketing programs making claims relative to carcass-quality characteristics. These "quality" claims are specified by the originator of the program and go above and beyond the day-to-day grading activities performed in the plants by USDA graders. As of this writing, 43 beef carcass certification programs exist (USDA, AMS, LSG, 2006), each varying in scope for live animal requirements, quality factors, yield factors, and other carcass characteristics. Process Verified programs, on the other hand, enable suppliers of products and/or services to "assure customers of their ability to provide consistent quality products or services" (USDA, AMS, ARCB, 2006). These programs are limited to specific process-verified points documented by a quality-management system, as defined by the supplier. Process Verified programs use the International Organization of Standardization's ISO 9000 as a basis for process documentation and review or audit practices. Such programs allow for suppliers to make marketing claims relative to their process-verified points relating to issues such as age, source, feeding, and production-processing claims.

SUMMARY

Quality is best defined by today's consumers' wants and desires for taste (discussed in Chapter 7, Sensory Attributes and Quality), safety (discussed in Part I), and preparation time/convenience; three of the top six factors consumers rate as "very important" for food selection (VanOverbeke, 2006). Everyone in the beef industry, from producers to processors and restaurateurs and retailers, has an obligation to meet these and other consumer demands that define quality.

The long-term advantage of the U.S. beef industry rests in its ability to meet consumer demand by providing quality products that meet or exceed consumer expectations. While there are many examples of success, participants in the beef industry must undertake the challenging process of incorporating a quality philosophy into their day-to-day decisions, production systems, and marketing choices. Such a transformation will not be easy but the long-term rewards to the industry and its customers justify the commitment.

LITERATURE CITED

Cattle-Fax. 2005. Beef Export Update. Cattle-Fax Resources, Centennial, CO.

Deming, W. Edwards. 1994. *The New Economics for Industry, Government, Education,* 2nd Edition. Massachusetts Institute of Technology, Cambridge, MA.

Food and Agricultural Organization of the United Nations and World Health Organization. 2005. *Understanding the Codex Alimenatius.* Rome, Italy.

Hurburgh, C.R. and J.D. Lawrence. 2002. The need for quality management systems in agriculture. *Resource Magazine* (December).

International Beef Quality Audit. 2004. Executive Summary. United States Meat Export Federation. Denver, CO.

Lawrence, J. 2005. Cattle producers need a Quality System Assessment for Japan. *Iowa Ag Review* 11: 3.

National Cattlemen's Beef Association. 1991. Final report of the National Beef Quality Audit—1991. G.C. Smith, Ed. Englewood, CO.

National Cattlemen's Beef Association. 1995. Final report of the National Beef Quality Audit—1995. G.C. Smith, Ed. Englewood, CO.

National Cattlemen's Beef Association. 2000. Final report of the National Beef Quality Audit—2000. G.C. Smith, Ed. Englewood, CO.

Peters, T. 1987. *Thriving on Chaos: Handbook for a Management Revolution.* Alfred A. Knopf, New York.

Smith, G.C., J.W. Savell, J.B. Morgan, and T.E. Lawrence. 2006. National Beef Quality Audit—2005: A new benchmark for the U.S. beef industry "Staying on Track." National Cattlemen's Beef Association, Centennial, CO.

Smith, G.C., J.W. Savell, J.B. Morgan, and T.H. Montgomery. 2000. Improving the quality, consistency, competitiveness and market-share of fed-beef—The final report of the third blueprint for total quality management in the fed-beef (slaughter steer/heifer) industry, National Beef Quality Audit—2000. National Cattlemen's Beef Association, Centennials, CO.

USDA, AMS, ARCB. 2006. USDA Process Verified Program. United States Department of Agriculture, Washington, DC. Available at processverified.usda.gov. Accessed on February 20, 2006.

USDA, AMS, LSG. 2006. Comparison of certified beef programs. United States Department of Agriculture, Washington, DC. Available at www.ams.usda.gov/lsg/certprog/speccomp.pdf. Accessed on February 20, 2006.

VanOverbeke, D.L. 2006. Consumer trends vs. beef production. Prepared for the Ontario Cattle Feeders Annual Convention. London, Ontario, Canada.

Walton, M. 1986. *The Deming Management Method.* The Berkley Publishing Group, New York.

Chapter 5

Preharvest Beef Quality

Robert A. Smith

HISTORY OF PREHARVEST BQA

Cattlemen have established a long tradition of producing quality beef. Advancing technologies and knowledge about the advantages and risks of using these technologies have brought about many changes in production practices. Research and education have provided the cornerstone for improvements in production practices on the farm, ranch, and feedlot that have resulted in improved beef quality and safety.

In 1982, the United States Department of Agriculture-Food Safety Inspection Service (USDA-FSIS) began working with the beef industry to develop the Preharvest Beef Production Program. To minimize additional governmental regulation, the beef industry developed an industry-driven program that was later named Beef Quality Assurance (BQA). The BQA program evaluated production practices at three large beef feedlots in Oklahoma and Kansas, and assessed the risk of chemical residues (Smith, Griffin, and Dargatz, 1997).

During the next two to three years, production practices were evaluated and modified at these three feedlots, and in 1985 they were certified by USDA-FSIS as "verified production control" feedlots. The lessons learned during this period of time demonstrated that industry-driven programs can be very effective, and that producers and governmental regulatory agencies can work together to solve problems and improve beef safety. This concept developed the framework for

Handbook of Beef Safety and Quality

doi:10.1300/5640_05

BQA programs that were later spearheaded by the National Cattlemen's Beef Association (NCBA) and other allied organizations.

Injection-site blemishes or lesions in muscle tissue were identified in the late 1980s. The NCBA's Beef Quality Assurance Advisory Board developed a plan to identify the prevalence of the lesions and research was conducted to determine the cause(s). Research grants were awarded to audit the prevalence of injection lesions in the top butt three times each year, establishing a benchmark and measurements of progress. Other studies defined the causes of injection-site lesions, and a massive program was initiated to educate producers, veterinarians, university faculty, and the drug and biological industry about the scope of the problem and how lesions could be avoided.

Ongoing audits showed that people who administered animal health products got the message and committed themselves to solving the problem. The dramatic reduction in injection-site lesions again clearly demonstrated that industry-driven programs can work, and serve as a model for follow-up efforts to improve beef quality and safety. The high-profile campaign to reduce injection-site lesions in beef put the BQA program on the map.

Research efforts were expanded to monitor quality defects of beef. Periodic fed cattle audits and cow audits were conducted to determine the incidence of quality defects, bruises, neoplasia, liver condemnation, hide damage, and so on, in order to make recommendations to alter management and, therefore, improve beef quality. These findings also offer guidance to producers to capture missed opportunities to optimize harvest value of their livestock.

Most recently, cattle care and handling guidelines have been incorporated into BQA programs. Animal care had previously been addressed on several fronts during the past several years, but guidelines were not comprehensive and detailed. Cattle care and handling guidelines developed by cattlemen have been endorsed by major veterinary organizations and groups representing grocers and restaurants.

A significant key to success of the BQA programs is the extensive networking between the NCBA, state beef organizations and beef councils, extension personnel, veterinary groups and experts from a wide range of fields. As one longtime member of the BQA Advisory Board has said many times, "There are no most valuable players when it comes to BQA" (Griffin, personal communication). Everyone's commitment and involvement is important. BQA has been successful because it is not a "top-down" program—basic programs are developed and provided through research, but through the BQA network, the final programs are tailored at the state and local level. Buy-in is essential!

Throughout its history, the goal of BQA has been simple—improve quality of beef in order to provide the consumers with what they want. If accomplished, this improves consumer demand and increases the likelihood of profit for beef producers.

RESIDUE AVOIDANCE

Pharmaceuticals, biologicals, and chemicals are necessary for prevention and treatment of disease in beef cattle. Drugs and vaccines have greatly improved the health and productivity of both humans and animals. Use of these products in food animals brings the solemn responsibility to ensure that their usage is safe for consumers and the animals themselves. Various government regulations, educational programs conducted by agricultural and veterinary groups, and willingness of cattle producers to take the production of safe beef seriously have reduced the risk of violative residues in cattle resulting from the use of pharmaceuticals, vaccines, and chemicals to negligible levels.

Regulations

The first Federal Food and Drug Act in the United States was passed in 1906, which banned adulterated or misbranded drugs from interstate commerce (Sundlof, 2001). The original law did not, however, ban false therapeutic claims. Numerous other laws have been passed since that time to regulate drugs, vaccines, and chemicals, but a detailed discussion is beyond the scope of this chapter.

It is important, however, for beef cattle producers to have a basic understanding of laws and regulations that provide the framework for the development, approval, distribution, and use of drugs and chemicals in livestock. The landmark Federal Food, Drug, and Cosmetic Act (FFD&C) was passed in 1938. This act prohibited interstate commerce of animal or human drugs unless they had been properly tested for safety when used under conditions prescribed on the label—basically it became illegal to ship drugs across state lines that were not shown to be safe. This act also dictated labeling requirements, but did not require proof of effectiveness (Sundlof, 2001).

In the 1950s and 1960s, amendments to the FFD&C Act had a significant effect on animal drugs—most notable were amendments led by Congressman Delaney and the Kefauver-Harris Amendments. As a result of these amendments, sponsors (manufacturers) of drugs intended for use in food animals had to demonstrate the safety to humans of any residue present in food products from tested animals. It also became necessary for sponsors of drugs to demonstrate that drugs or drug residues will not cause cancer in animals or humans. Further amendments required manufacturers to provide evidence of effectiveness of new drugs, as well as their safety, in order to gain Food and Drug Administration (FDA) approval (Sundlof, 2001).

In 1994, Congress passed the Animal Medicinal Drug Use Clarification Act, commonly referred to as AMDUCA (CVM Update, 1996). This law decriminalized extra-label use of drugs by veterinarians (Sundlof, 2001), but not laymen. Until passage of this act, no provisions under the law existed to allow veterinarians to use drugs other than as specified on the label—in other words, no deviation from the labeled dosage, route of administration, species, or indication (condition treated) was allowed. Prior to this time, the FDA had exercised regulatory discretion and allowed most extra-label usage of drugs to continue, but this was technically illegal. The practice of extra-label usage existed because there are many conditions for which there is no labeled product for treatment, and without treatment unnecessary death or suffering of animals would occur.

AMDUCA allows FDA-approved animal and human drugs to be used extra-label when no effective labeled product exists, but extra-label use of drugs is permitted only when a valid veterinarian–client–patient relationship exists. Extraordinary care must be taken to ensure that use of products in an extra-label manner does not result in violative drug residues. It is important for producers to understand that extra-label drug usage is allowed only when the health of the animal is threatened, or when death or suffering is likely to occur without treatment. Drugs cannot be used extra-label for production purposes, such as extra-label use of growth implants (Fajt, 2003), or for economic reasons. The law specifically prohibits the extra-label use of some compounds in food-producing animals under any circumstances. These çompounds are fluoruquinolones, glycopeptides, furazolidone, nitrofurazone, chloramphenicol, dimetridazole, ipronidazole, other nitromidazoles, certain sulfonamide drugs in lactating dairy animals (check product labels), clenbuterol, and diethylstilbestrol (Fajt, 2003). In addition, extra-label use of any livestock feed additive is strictly prohibited. Obviously consumer confidence is greater when animal health products are used in concert with the approved labeling.

In the end, the purpose of laws and regulations governing the use of drugs and chemicals in livestock is to protect animal and human safety, and to provide assurance to producers that the products are safe and effective when used in their livestock as labeled. These controls and individual commitment to food safety by producers and veterinarians have resulted in negligible residue problems in beef.

Over-the-Counter Drugs and Prescription Drugs

Beef producers routinely use drugs to maintain health of their herds. Drugs are classified into one of two categories: "over the counter" (OTC) or "prescription." OTC drugs can be purchased and used without a veterinary prescription, but can only be legally used in strict accordance with the label—no deviation is allowed. These drugs are labeled for very specific and limited uses (Sundlof, 2001), and include such common products as penicillin and tetracyclines. Procaine penicillin G is a classic example of a drug unknowingly used extra-label by many producers. For cattle, it is labeled only for treatment of bacterial pneumonia (shipping fever) caused by *Pasteurella multocida* at a dosage of 1 ml per 100 pounds of body weight, administered intramuscularly (IM). If it is used to treat foot rot, for example, administered at a higher dosage or by another route of administration, this constitutes extra-label usage and is illegal for laymen. Changing the dosage or route of administration can result in violative drug residues when cattle are harvested. In our penicillin example, the approved preharvest withdrawal time is 10 days when used according to label instructions, but is much longer when a higher dosage is used or when it is given subcutaneously (SC).

Many drugs used to prevent and treat diseases in beef cattle are restricted to use by or on the order of a licensed veterinarian. These are commonly called "prescription" drugs and can be readily identified by the statement found on the label stating "Caution: Federal (USA) law restricts this drug to use by or on the order of a licensed veterinarian." In order to possess and use these products in cattle, producers must obtain a "prescription" or "drug purchase order" from a veterinarian—the term used mostly depends on state pharmacy laws. When dispensed or prescribed by the veterinarian, these products can be used only in accordance with label instructions unless the veterinarian has determined that extra-label usage is necessary.

It is important for producers to understand constraints (AMDUCA, 1994) placed on veterinarians to prescribe drugs or to advise extra-label usage. In order to do so, there must be a valid veterinarian–client–patient relationship (VCPR) in which:

1. The veterinarian has assumed the responsibility for making medical judgments regarding the health of (an) animal(s) and the need for medical treatment, and the client (the owner of the animal or animals or other caretaker) has agreed to follow the instruction of the veterinarian.
2. There is sufficient knowledge of the animal(s) by the veterinarian to initiate at least a general or preliminary diagnosis of the medical condition of the animal(s).
3. The practicing veterinarian is readily available for follow-up in case of adverse reactions or failure of the regimen of therapy. Such a relationship can exist only when the veterinarian has recently seen and is personally acquainted with the keeping and care of the animal(s) by virtue of examination of the animal(s), and/or by medically appropriate and timely visits to the premises where the animal(s) is (are) kept.

Requirements for Extra-Label Drug Usage in Beef Cattle

There are specific circumstances where extra-label usage is allowed under AMDUCA (1994):

1. There is no approved new animal drug that is labeled for the intended use and that contains the same active ingredient which is

in the required dosage form and concentration, except where a veterinarian finds, within the context of a valid veterinarian–client–patient relationship, that the approved new animal drug is clinically ineffective for its intended use.

2. Prior to prescribing or dispensing an approved new animal or human drug for an extra-label use in food animals, the veterinarian must:
   - Make a careful diagnosis and evaluation of the conditions for which the drug is to be used;
   - Establish a substantially extended withdrawal period prior to marketing of milk, meat, eggs, or other edible products supported by appropriate scientific information, if applicable;
   - Institute procedures to assure that the identity of the treated animal or animals is carefully maintained;
   - Take appropriate measures to assure that assigned time frames for withdrawal are met and no illegal drug residues occur in any food-producing animal subjected to extra-label treatment.

The following additional conditions must be met for a permitted extra-label use in food-producing animals of an approved human drug, or of an animal drug approved only for use in animals, not intended for human consumption:

1. Such use must be accomplished in accordance with an appropriate medical rationale.
2. If scientific information on the human food safety aspect of the use of the drug in food-producing animals is not available, the veterinarian must take appropriate measures to assure that the animal and its food products will not enter the human food supply.
3. Extra-label use of an approved human drug in a food-producing animal is not permitted under this part if an animal drug approved for use in food-producing animals can be used in an extra-label manner for the particular use.

Practical Management Steps for Residue Avoidance

Producers should read, understand, and follow label instructions before using any animal health product. As suggested previously, any usage that deviates from label instructions should be done only on the advice of a veterinarian. Of special significance is the effect of extra-label usage on the preharvest withdrawal time. Owners or managers must ensure that employees or family members who administer animal health products are properly trained on residue avoidance and other BQA procedures.

Most, but not all, animal health products have withdrawal times that must be observed prior to harvest. It is also imperative that producers understand that purchasers of livestock at auctions or private treaty might resell animals for harvest; therefore, the best practice is to withhold animals from marketing for any purposes until the withdrawal time has passed. If a preharvest withdrawal time for an animal health product is required, a statement will be found on the label stating "Warning(s): Animals intended for human consumption must not be slaughtered within xx days of last treatment," or "Warning(s): Discontinue treatment at least xx days prior to slaughter of cattle." Accurate record-keeping is vital for residue avoidance. The date, product used, dosage, and route of administration should be recorded, along with group or individual animal identification.

Group identification, such as pen, pasture, or lot, is acceptable when there is no mixing of cattle from other locations during the withdrawal period. It is appropriate when all animals are vaccinated, dewormed, or treated with a product requiring a preharvest withdrawal time. Each animal treated individually should be uniquely identified, such as with a numbered eartag. Other forms of individual identification are acceptable—tattoo, numbered freeze brand, and electronic identification.

Products should be administered at the dosage and by the route of administration—IM, SC, IV, oral, or topical—as stated on the label. Many product labels also specify the maximum volume of drug to be administered at each injection site. Exceeding the label dosage, volume at each injection site, using a different route to deliver the drug, or administering more than the recommended number of treatments may result in drug residues beyond the withdrawal time. When drugs have been used extra-label, it is the responsibility of the attending veterinarian to determine the extended withdrawal time, which the producer must adhere to. The Food Animal Residue Avoidance Databank (FARAD) is a common source of information for veterinarians to make decisions about the appropriate withdrawal times for products used extra-label (Riviere and Sundlof, 2001).

At market time, animal health records must be carefully checked to ensure that all animals have cleared the required preharvest withdrawal time. Errors are minimized if more than one person reviews the records. The signature of the person who did the withdrawal report and the date the report was done establishes accountability. This practice is highly recommended in larger operations where designated employee(s) prepare(s) the withdrawal report.

By understanding laws and regulations governing drug usage in cattle, having a residue avoidance program in place, training employees on all facets of the program, and working with a veterinarian within a valid VCPR, violative drug residues should not be an issue in beef production. Consumers expect no less.

INJECTION-SITE LESIONS

Injection-site lesions are also commonly called injection-site scars, blemishes, or defects. These result from administration of injectable animal health products. In muscle, visible scars can be seen at the site of injection, resulting in trim loss. Research has also demonstrated a loss of tenderness as far as three inches (7.62 cm) away from the visible lesion, even though the muscle looked grossly normal (George, Cowman et al., 1996).

Until the 1990s, injection of vaccine and medication into muscles of the top butt and rounds was commonplace. This resulted in significant trim loss and decreased beef quality. Many farmers and ranchers assumed that muscle damage caused by injection of vaccines and drugs would "heal" by the time the animal was harvested. However, research by George et al. (1995) showed that IM administration of clostridial vaccine and certain antibiotics to calves caused damage so severe that lesions persisted until harvest 7.5 to 12 months later (Figure 5.1).

There is evidence that additional vaccines and pharmaceutical compounds cause injection-site lesions when injected into muscle tissue, including sterile saline, modified live virus vaccines, inactivated (killed) virus vaccines, clostridial bacterins, vitamins, and aqueous, macrolide and oxytetracycline antibiotics (George, Ames, et al.,

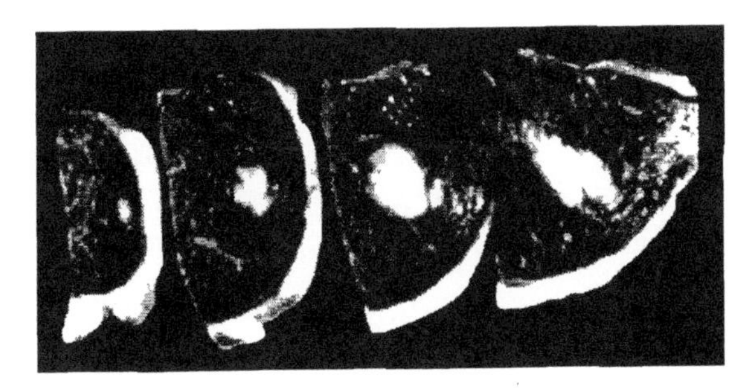

FIGURE 5.1. Injection-site lesions resulting from injection of clostridial bacterin-toxoid at 50 days of age; steer was harvested approximately one year after vaccination. (Courtesy of Dr. Gary Cowman, National Cattlemen's Beef Association. Used by permission.)

1996). The bottom-line message is clear—any product injected into muscle tissue can cause injection lesions, trim loss, and decreased beef quality, particularly loss of tenderness.

Intramuscular versus Subcutaneous Administration

The approved route of administration for animal health products is stated on the label. When products are labeled for either IM or SC administration, the SC route should be used to avoid damage to edible tissue. Similar products produced by different manufacturers may have different labeled routes of administration. If the products are otherwise equal, the one labeled for SC administration should be used. When possible, IM injections should be avoided. Deviation from the stated route of administration is extra-label.

IM and SC Injections

All injections should be given in front of the shoulders—never in the top butt (rump) or round (leg). Recommended location for administration of injectable animal health products is illustrated in Figure 5.2. It is important to avoid giving injections over the shoulder, at the junction of the neck and shoulder, crest of the neck, or into or over the cervical vertebral column located distal to the approved "injection triangle." This practice should be followed for all classes and ages of

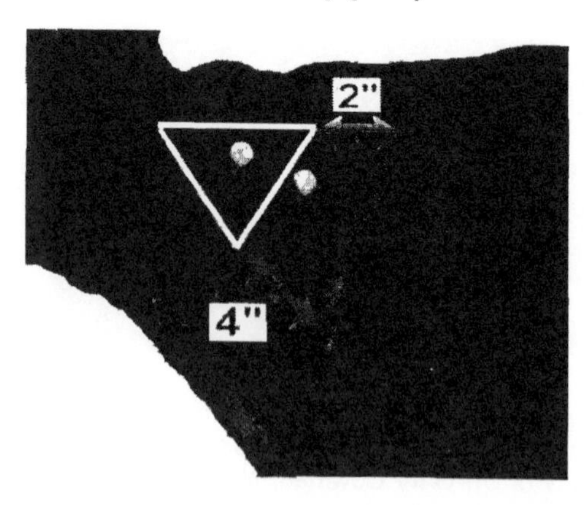

FIGURE 5.2. Avoid intramuscular injections 2 to 4 inches forward of the slope of the shoulder to minimize injection-site defects in the underlying shoulder roll. (Courtesy of Dr. Dee Griffin, University of Nebraska. Used by permission.)

beef cattle. "Tenting" of the skin, or pulling the skin away from underlying tissues as an injection is given, can result in more accurate administration of SC products (Figure 5.3); however, a one-handed technique can be effective if the correct angle of needle insertion is employed.

Selection of the right needle for the product to be administered is an important part of BQA. Needles should be of high quality, and should be replaced when they become dull, burred, or after a soiled animal has been injected. Frequency of needle change is also influenced by the risk of transmitting blood-borne diseases, such as anaplasmosis. A 16-gauge × 1/2" to 3/4" needle should be used to inject products SC, although the manufacturer of tilmicosin recommends a maximum needle length of 5/8" because of human safety concerns. A 16-gauge × 1" needle is typically recommended for IM injections; however, large, thick-hided bulls or adult cows may require a 1.5" needle. If a needle bends, it should never be straightened and reused. On rare occasions a needle can break off in muscle tissue, posing a risk to the consumer. Under no circumstances should an animal with a broken needle in it be allowed to enter marketing channels (Smith et al., 2001).

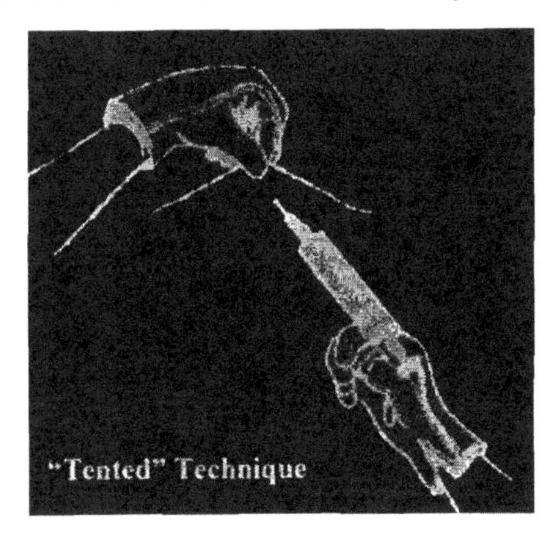

FIGURE 5.3. Pulling the skin away from underlying tissues, or "tenting" the skin, can make subcutaneous administration of animal health products more accurate. (Courtesy of Dr. Gary Cowman, National Cattlemen's Beef Association. Used by permission.)

Volume of Product in Each Injection Site

BQA guidelines state that no more than 10 ml should be administered IM at each injection site. The maximum volume of drug administered SC at each site varies from product to product, depending on the label. In general, the labeled volume per SC site ranges from 10 to 20 ml. Exceeding the recommended volume at each site can result in violative drug residues or poor absorption. Multiple injection sites should be separated by at least a hand's width.

Animal Restraint

Proper restraint of cattle is essential to give injections properly and to ensure safety of personnel giving the injections. Injections or "shots" cause pain, and when animals are improperly restrained they try to escape. Excessive animal movement results in unpredicted movement of the needle in the SC area or in the muscle, which can cause physical damage to the tissue that heals by scarring.

Catching cattle in a chute with a headgate and side squeeze minimizes movement. Animals should be caught in a chute so that ample neck area is available for SC and IM injections; generally, catching the animal with the headgate just behind the ears accomplishes this. This method also restricts movement of the head, making procedures such as implanting, tagging, and administration of oral predications easier and safer.

Record Keeping and Documentation

Accurate records must be kept for all preventive or therapeutic treatments. This is especially important to avoid marketing of animals that have not cleared their mandated withdrawal times. Alliances and source/process verification programs have gained in popularity. Animal identification and documentation of what was done, when it was done, and who did it are key components of many of these programs.

CARE AND HANDLING OF BEEF CATTLE

Proper care and handling of livestock is an integral component of BQA. Sound animal husbandry practices are known to impact the health and productivity of beef herds. Societal concerns about how beef and other animal protein sources are produced have led to development of care and handling guidelines by the various commodity groups. Many national, state, and local cattlemen's organizations have incorporated cattle care into their BQA programs. Common components of cattle care guidelines include nutrition and feeding, disease prevention practices, cattle handling, marketing, management of downer animals, and euthanasia. Much of the information presented here is the product of the National Cattlemen's Beef Association's *Guidelines for Care and Handling of Beef Cattle,* which was developed by the NCBA Cattle Care Working Group, and approved by the NCBA Board of Directors.

Feeding and Nutrition

Nutrition requirements of beef cattle vary according to age, sex, weight, breed, or biological type, weather, body condition, and stage of production. Diets for all classes of beef cattle, grazing, or feeding, should meet the recommendations of the National Research Council (NRC) and/or recommendations of a feeding consultant. Nutrient requirements for all classes of beef cattle are found in the NRC *Nutrient Requirements of Beef Cattle* (1996).

Beef cows must be fed to sustain health and reproduction. Body condition scoring of beef cows is a scientifically approved method to assess nutritional status. Body condition scores (BCS) range from 1 (emaciated) to 9 (obese). A BCS score of 4 to 6 is most desirable for health and production. During periods of prolonged drought and widespread shortages of hay and other feedstuffs, the average BCS of cows within a herd may temporarily decline. A BCS of 2 or under is not acceptable and immediate corrective action should be taken.

On growing forages, stocking rates for stocker calves should be established to meet production goals for growth and performance. When pastures are dormant, stocker cattle should be supplemented to meet growth requirements as established by the NRC and targeted production goals of the operation.

Feedlot cattle consume higher-energy diets; therefore a careful feeding plan must be developed and monitored to minimize digestive upsets. An ongoing relationship with a nutritionist, and proper training of personnel responsible for manufacturing, delivering, and making decisions about ration transition are essential. Increases in bloat, acidosis, or laminitis above targeted levels require immediate investigation in order to minimize losses or compromise in welfare of the cattle. Cattle with severe laminitis, when the hoof length is double its normal length, should be provided appropriate care; early marketing should be considered.

Under some circumstances, such as drought, frosts, and floods, some commonly used feedstuffs can become toxic and cause severe illness or death of cattle. Examples of toxins include nitrates, prussic acid, and mycotoxins. When environmental and plant conditions are favorable for toxin formation, feedstuffs should be analyzed by a laboratory before being fed. Producers should also become familiar with potential micronutrient deficiencies or excesses, and use appropriately formulated supplements.

Cattle must have access to an adequate water supply. Estimated water requirements for all classes of beef cattle in various production settings are described in the NRC *Nutrient Requirements of Beef Cattle* (1996).

Disease Prevention and Health Care

Like other species, cattle are susceptible to infectious diseases, metabolic disorders, toxins, parasites, neoplasia, and injury. Health-management plans can reduce economic loss, improve herd productivity, and improve animal welfare.

A detailed discussion of health care practices is beyond the purpose and scope of this chapter. However, in general terms, cows and bulls should be vaccinated against respiratory, reproduction, or other diseases based on risk assessment and efficacy of available vaccines. Parasite control should also be based on risk assessment, utilizing such tools as fecal examination and BCS to make the final decision.

During calving season, cows should be checked regularly for calving difficulties. First-calf heifers may require more frequent observation and care. Cows and heifers should be allowed to calve on open pasture unless weather or possibility of calving difficulty dictates otherwise. If a calving barn is used, ample room for the cow/heifer to deliver her calf naturally must be provided. Fresh bedding should be provided and changed frequently to avoid soiling and disease transmission.

Cows with mild lameness, early eye problems such as ocular neoplasia, mastitis, or loss of body condition should be examined to determine well-being and, in some cases, should be promptly marketed.

Procedures such as castration and dehorning of calves are done for the protection of the animal, other cattle in the herd, and people who handle the cattle. Early castration improves animal performance and reduces health complications. Castration prior to 120 days of age or when calves weigh less than 500 pounds is strongly recommended (FASS, 1999). Acceptable reasons to delay castration are if bull calves are being considered as seedstock or to be finished as intact bulls. Producers should vaccinate against tetanus when bands are used for castration.

When horns are present, it is recommended that calves be dehorned prior to 120 days of age. Dehorning should be done before the diameter of the horn base grows to one inch in diameter or more. It is strongly recommended that a local anesthetic, called a cornual nerve block, be used if the horn base is one inch or more in diameter. Tipping of horns (removing the tip only) can be done with little impact on the well-being of individual animals (FASS, 1999).

Weaning is less stressful when calves are castrated and dehorned early in life, vaccinated against respiratory diseases prior to weaning, and provided proper preweaning nutrition. Calves should be vaccinated and treated for parasites based upon risk assessment and the efficacy of available animal health products. Stress is decreased if calves are weaned for approximately 45 days before shipment to a stocker operation or feedyard.

Health problems in stocker operations and feedlot are typically greater than on the ranch, particularly Bovine Respiratory Disease (BRD). Weaning, commingling, transport, and dietary changes are major stressors. All incoming stocker and feeder cattle should be vaccinated against BRD. Stocker cattle intended to graze rangeland or pasture should be vaccinated against clostridial diseases. The use of other vaccines and parasite control should be based on risk assessment and efficacy of available animal health products.

Pregnancy in immature heifers can result in calving difficulties and subsequent trauma to the birth canal, paralysis, or death of the heifer. For these reasons it is often more humane to abort pregnant heifers in stocker and feedlot operations. This should be done under the direction of a veterinarian. If heifers in the feedyard or a stocker operation deliver a full-term, healthy calf, it should be allowed to nurse to obtain colostrum. At all times, these calves must be handled humanely and provided proper nutrition. Compromised calves or fetuses should be promptly euthanized and disposed of according to local regulations.

Cattle Handling

Cattle are gathered to perform routine husbandry procedures, such as veterinary care, weighing, sorting, weaning, and transportation to and from pastures, feedlots, and livestock markets. Handling procedures must be safe for the cattle and caretakers, and cause as little stress as possible. Facilities should be designed and constructed to take advantage of cattle's natural instincts. They do not have to be elaborate or expensive, but should be designed in a manner that does not impede cattle movement. Slippery surfaces, especially where cattle enter a single file alley leading to a chute or exiting a chute, should be avoided to minimize slipping, falling, and toe abscesses. Under most conditions, no more than 2 percent of the animals should fall outside the chute (Grandin, 1998). A level of more than 2 percent dictates a review of the process, including asking questions such as the following: Is this a cattle temperament issue? Has something in the handling area changed that is affecting cattle behavior? Quiet handling greatly reduces the risk of slipping and falling. Grooved concrete, rubber mats, or deep sand can be used to improve traction.

All personnel working with cattle should be trained on cattle behavior and handling. Understanding cattle's flight zone and point of balance (Grandin, 1998) is critical for effective, safe movement of cattle.

Cattle handlers should take advantage of cattle's flight zone and point of balance to move them. For safety and welfare reasons, the use of electric prods should be minimized. Nonelectric driving aids, such as plastic paddles, sorting sticks, flags, or streamers (affixed to long handles) can be effectively used to quietly guide and turn animals. When cattle continuously balk, cattle handlers should investigate and correct the reason rather than resort to overuse of electric prods. Under most conditions, 90 percent or more of cattle should flow through cattle handling systems without the use of electric prods (Grandin, 1998). When cattle prods must be used, avoid contact with the eyes, rectum, genitalia, and udder. Driving aids powered by AC current should never be used unless manufactured and labeled specifically for that purpose. Voltage must be regulated to less than 50 volts.

Some cattle are naturally more prone to vocalize, but if more than 5 percent of cattle vocalize (after being squeezed but prior to procedures being preformed), it may be an indication that chute operations should be evaluated (Grandin, 1998). Key questions to ask include the following: Is this a cattle temperament issue or a result of prior handling? Are chute pressures and catching methods appropriate, or should they be reevaluated? If more than 25 percent of cattle jump or run out of the chute there should be a review of the situation and questions asked such as the following: Is this a result from cattle temperament or prior handling practices? Was the chute operating properly, etc.? Evaluate handling procedures to determine whether practices need to be improved or whether the problem is cattle temperament.

Marketing Cattle

The overwhelming majority of cattle are marketed in good health and physical condition. Some compromised cattle should not enter intermediate marketing channels because of animal welfare concerns. Instead, these cattle should be sold directly to a processing plant or euthanized, depending upon the severity of the condition, processing plant policy, and state or USDA regulations.

Non-Ambulatory (Downer) Cattle

Cattle can become downers for several reasons, including injury, severe disease, and chronic emaciation. With proper care, many of these will recover and become productive animals. It is the responsibility of livestock owners and caretakers to make every effort to provide proper care for non-ambulatory livestock. Physical management of downer cattle presents challenges, as they may weigh over 1,000 pounds.

A prompt diagnosis should be made to determine whether the animal should be humanely euthanized or receive additional care. Feed and water must be provided to non-ambulatory cattle at least once daily. Downer animals must be moved very carefully to avoid compromising animal welfare. Dragging downer animals is unacceptable. Likewise, animals should not be lifted with chains onto transportation conveyances. Acceptable methods of transporting downers include a sled, low-boy trailer, or in the bucket of a loader. Animals should not be "scooped" into the bucket, but rather should be humanely rolled into the bucket by caretakers.

When treatment is attempted, cattle unable to sit up unaided (i.e., lying flat on their side) and refusing to eat or drink should be humanely euthanized within 24 to 26 hours of initial onset unless a veterinarian recommends otherwise. Signs of a more favorable prognosis include the ability to sit up unaided, eating, and drinking. It is acceptable to allow more time for recovery for these animals, provided they are offered water, feed, and the weather is moderate to good. If weather creates inhumane conditions or the animal's condition deteriorates, it should be humanely euthanized.

Cattle that are non-ambulatory must not be sent to a livestock market or to a processing facility. If the prognosis is unfavorable or the animal has not responded to veterinary care, it should be humanely euthanized.

Euthanasia

Euthanasia is humane death occurring without pain and suffering. The decision to euthanize an animal should consider the animal's welfare. The producer will most likely perform on-farm euthanasia because a veterinarian may not be immediately available to perform the service. The person performing the procedure should be knowledgeable of the available methods and have the necessary skill to safely perform humane euthanasia; if not, a veterinarian must be contacted. When euthanasia is necessary, an excellent reference is the *Practical Euthanasia of Cattle* guidelines developed and published by the American Association of Bovine Practitioners.

Reasons for euthanasia include the following:

- Severe emaciation, weak cattle that are non-ambulatory or at risk of becoming downers;
- Downer cattle that will not sit up, refuse to eat or drink, have not responded to therapy and have been down for 36 hours or more;
- Rapid deterioration of a medical condition for which therapies have been unsuccessful;
- Severe, debilitating pain;
- Compound (open) fracture;
- Spinal injury;
- Central nervous system disease;
- Multiple joint infections with chronic weight loss.

Feedlot Heat-Stress Procedures

Cattle deaths during periods of high heat and humidity can be an economic disaster for the owners, and the effect on animal welfare tends to generate negative news stories. While producers cannot change the weather, there are steps that can be taken to mitigate the impact of high heat and humidity.

Adequate water should always be available. If possible, avoid handling cattle when the risk of heat stress is high. The final decision must consider temperature, humidity, wind speed, phenotype, and cattle acclimation. If cattle must be handled, a general rule is to work them before the Temperature Humidity Index (THI) reaches 84. As an example, when the temperature is 98°F and the humidity is 30 percent, the THI is 83. At a constant temperature, the THI increases as the relative humidity increases. Each one mile per hour increase in wind speed decreases the THI by approximately one. More information can be found in NebGuide G00-1409-A (www.gpvec.unl.edu).

Cattle more prone to heat stress should be worked earlier in the day. For example, larger cattle, such as those being implanted, should be processed during the cooler times of the day. The time cattle spend in handling facilities where heat stress may be more significant should be limited. When crowded into small alleys or pens awaiting processing, or when trapped behind solid fences found in processing areas, air movement is limited and heat stress can result.

When heat stress is extreme or life-threatening, such things as creating water holes, wetting the pen surface, and wetting cattle with sprinklers or hoses can reduce heat stress and minimize losses.

Emergency Procedures

A plan to ensure the welfare of the animals when unforeseen emergencies occur should be in place. Post names and telephone numbers of the producer or management, veterinarian, equipment suppliers, and the fire and police departments near telephones, along with directions to the cattle operation, including road names and numbers. The person in charge should review possible emergencies that might arise and review these plans with other employees, so that everyone is familiar with the appropriate emergency response.

RELATIONSHIP OF DISEASE TO PERFORMANCE AND BEEF QUALITY

In recent years, several studies have demonstrated the adverse effects of sickness, especially Bovine Respiratory Disease (BRD), on cattle performance and carcass quality. Healthy herds are generally more productive, and health maintenance programs can improve performance and carcass quality, as well as address animal welfare issues. It is not the intent of this chapter to discuss specific herd health programs, but rather to illustrate their importance. Producers should develop herd health plans with their veterinarian, and ensure that nutritional programs also address health, reproduction, and performance goals.

BRD and Feedlot Performance

Many studies have demonstrated that cattle that have experienced BRD have reduced daily gain compared to those remaining healthy. These differences in average daily gain (ADG) can persist to harvest time. In a Canadian trial, calves treated for BRD early in the feeding period gained 0.14 pound per day less than those not treated (Bateman et al., 1990). Calves never treated gained as much as 0.56 pound per day more than treated calves in the Texas A&M Ranch to Rail studies (1992-1993 to 1999-2000).

The presence of lung lesions caused by BRD is also associated with reduced ADG. Lungs were examined at harvest in Nebraska and calves with pulmonary lesions gained 0.17 pound per days less than calves without lesions. The feeding period was 273 days (Wittum et al., 1995). Oklahoma researchers (Gardner et al., 1999) reported that steers without pulmonary lesions at harvest gained 3.48 pounds per day during a 150-day feeding period. Steers with inactive lesions gained 3.15 pounds per day, while those with "active" lesions gained only 2.57 pounds per day.

BRD, Lung Lesions, and Quality Grade

In the Texas A&M Ranch to Rail studies, there was a consistent difference in the percentage of cattle that graded USDA choice between those that had been treated for BRD during the feeding period and those remaining healthy. Cattle not sick and treated for BRD had as much as 17 percentage points more Choice carcasses than cattle that had been treated. In another study, 428-pound heifers were backgrounded for 42 days prior to feedlot entry. When harvested at the end of the finishing period, 66 percent of the heifers never treated for BRD during the backgrounding period graded Choice, while 59 percent of calves treated once and 41 percent of those treated more than once graded Choice, respectively (Stovall et al., 2000). This suggests that catabolic events, such as BRD, can have long-term effects on carcass quality and BQA. If the beef industry continues movement toward value-based marketing, the effect of BRD on carcass traits significantly adds to the economic liability posed by BRD.

OTHER BQA ISSUES

There are other important preharvest BQA issues that were not addressed in this chapter. Most notably, foodborne pathogens and evolving Process/Source Verification, Beef Export Verification, and other programs aimed at specific markets. Foodborne pathogens are discussed elsewhere in this text.

Animal identification was not discussed primarily because the National Animal Identification System (NAIS) is under development, but far from complete. Needless to say, there will be a national animal identification system in place within the foreseeable future. Its primary purpose is to provide 48-hour traceback in case of an animal health emergency, such as the introduction of a foreign animal disease, but the system will also provide a vehicle for other programs that can benefit from animal ID.

THE FUTURE OF PREHARVEST BQA

Producers have a responsibility to produce safe food; market forces demand that beef be both safe and of high quality. Everyone in the chain of production must do his or her part; any breakdown in the system compromises quality and safety.

Alliances, branded beef programs, and animal identification all place more and more demands to produce a safe and quality product. The marketplace has taken on a more global look, and as a result international standards and expectations now influence BQA. To meet these expectations, preharvest BQA programs must be continually updated to reflect new findings, and aggressive educational campaigns targeted toward all phases of production must remain ongoing.

Research must provide the framework for updates to preharvest BQA. BQA programs that span on-farm production and management, packing houses, fabrication, wholesale and retail distribution, and food service facilities should all be based on HACCP concepts. Preharvest BQA will continue to be an integral link in the beef quality and safety chain.

LITERATURE CITED

AMDUCA. 1994. Extralabel drug use in animals. www.access.gpo.gov/su_docs/aces/aces140.html (search on extralabel). Accessed January 23, 2006.

American Association of Bovine Practitioners, 1999. Practicle Euthanasia of Cattle: Considerations for the producer, livestock market operator, livestock transporter, and veterinarian. Auburn, AL.

Bateman K.G., S.W. Martin, P.E. Shewen, and P.L. Menzies. 1990. An evaluation of antimicrobial therapy for undifferentiated bovine respiratory disease. *Can. Vet. J.* 31: 689-696.

CVM Update. 1996. FDA publishes final rule on extralabel drug use in animals (November 12, 1996). www.fda.gov/cvm/CVM_updates/amducaup.html. Accessed January 23, 2006.

Fajt, V.R. 2003. Regulatory considerations in the United States. *Vet. Clin. North Am: Food Anim. Pract.* 19 (3): 695-705.

FASS. 1999. Guidelines for beef cattle husbandry. In: *Guide for the Care and Use of Agricultural Animals in Agricultural Research and Teaching.* 1st revised. Federation of Animal Science Societies, Savoy, IL, pp. 33-34.

Gardner, B.A., H.G. Dolezal, L.K. Bryant, F.N. Owens, and R.A. Smith. 1999. Health of finishing steers: Effects on performance, carcass traits, and meat tenderness. *J. Anim. Sci.* 77: 3168-3175.

George, M.H., R.P. Ames, R.G. Glock, M.T. Smith, J.D. Tatum, K.E. Belk, and G.C. Smith. 1996. Incidence, severity, amount of tissue affected and effect on histology, chemistry, and tenderness of injection-site lesions in beef cuts from calves administered a control compound or one of seven chemical compounds. Final Report to the National Cattlemen's Beef Association, Dept. of Animal Sciences, Colorado State University, Fort Collins, CO.

George, M.H., G.L. Cowman, J.D. Tatum, and G.C. Smith. 1996. Incidence and sensory evaluation of injection site lesions in beef top sirloin butts. *J. Anim. Sci.* 74 (9): 2095-2103.

George, M.H., P.E. Heinrich, D.R. Dexter, J.B. Morgan, K.G. Odde, R.D. Glock, J.D. Tatum, G.L. Cowman, and G.C. Smith. 1995. Injection-site lesions in carcasses of cattle receiving injections at branding and at weaning. *J. Anim. Sci.* 73: 3235-3240.

Grandin, T. 1998. Handling methods and facilities to reduce stress on cattle. *Vet. Clin. North Am: Food Anim. Pract.* 14 (2): 325-342.

National Research Council. 1996. *Nutrient Requirements of Beef Cattle,* 7th revised. National Academy Press, Washington, DC.

Riviere, J.E., and S.F. Sundlof. 2001. Chemical residues in tissues of food animals. In: *Vet. Pharm. Ther.*, 8th ed. H.R. Adams, ed. Iowa State University Press, Ames, IA, pp. 1166-1174.

Smith, R.A., D.D. Griffin, and D.A. Dargatz. 1997. The risks and prevention of contamination of beef feedlot cattle: The perspective of the United States of America. *Sci. .Tech. Rev.* (OIE). 16 (2): 359-368.

Smith, R.A., G.L. Stokka, O.M. Radostits, and D.D. Griffin. 2001. Health and production management in beef feedlots. In: *Herd Health: Food Animal Production Medicine,* 3rd ed. O.M. Radostits, ed. W.B. Saunders Co., Philadelphia, PA, pp. 617-621.

Stovall, T.C., D.R. Gill, R.A. Smith, and R.L. Ball. 2000. Impact of bovine respiratory disease during the receiving period on feedlot performance and carcass traits. Oklahoma State University Animal Science Research Report, pp. 82-86.

Sundlof, S.F. 2001. Legal control of veterinary drugs. In: *Veterinary Pharmacology and Therapeutics,* 8th ed. H.R. Adams, ed. Iowa State University Press, Ames, IA, pp. 1149-1156.

Texas A&M Ranch to Rail Summary Reports. 1992-2000. Texas A&M University, College Station, TX.

Wittum, T.E., N.E. Woolen, L.J. Perino, and E.T. Littledike. 1995. Relationship between treatment for respiratory disease, lesions at slaughter and rate of gain in feedlot cattle. *J. Anim. Sci.* 73 (Suppl I): 238 (abstr).

Chapter 6

Beef Carcass Quality

Jeff W. Savell
Carrie L. Adams Mason
F. Danielle Espitia
Diana Huerta-Montauti
Kristin L. Voges

INTRODUCTION

Beef quality encompasses many different organoleptic characteristics, but is primarily defined by three associated sensory factors: tenderness, juiciness, and flavor. The importance of beef quality stems from the purchasing power of consumers and their perception of meat quality. Initiating consumer beef purchases, allowing them to recur and ultimately retaining them, are entirely affected by the quality of meat offered to the consumer.

Function of Quality Grades

The purpose of grading beef carcasses is to identify specific traits and to utilize these to stratify a heterogeneous supply into homogenous groups. More specifically, quality grading is used as an estimation of palatability, or the satisfaction in taste and texture properties of meat. Marbling and maturity, collectively, are evaluated to assess and determine quality grades of beef carcasses. Both marbling and maturity are very influential aspects in terms of meat quality and are recognized as contributing factors to all aspects of palatability. Re-

Handbook of Beef Safety and Quality

doi:10.1300/5640_06

search has shown that as quality grade increases, variability decreases, and overall average desirability increases (Smith et al., 1984).

Background/History of Carcass Quality and Grading

Basis for the historical material to follow comes from the reviews by Briskey and Bray (1964) and Harris, Cross, and Savell (1990). In the early 1900s, the need for beef carcass grading standards was first suggested in "Market Classes and Grades of Cattle with Suggestions for Interpreting Market Quotations" by Herbert Mumford, a professor at the University of Illinois, to facilitate market reporting and to relay market and feedlot requirements to beef cattle feeders and breeders. Seven market grades—Prime, Choice, Good, Medium, Common, Cutter, and Canner—were defined with descriptions and pictures. These inaugural grades ultimately led to the beef grades currently used. Both livestock and meat industries, as well as consumers, were insistent on developing uniform standards for grading livestock and meat. This persistence led to the formation of the United States Department of Agriculture (USDA) by Congress in 1914 for the purpose of tending to agriculture marketing and establishing the Office of Markets and Rural Organization. Two years later, Congress created the National Livestock Marketing News Service and, as a result, a market classification system was essential for accurate market reporting. In 1916, USDA started formulating grade standards with uniform classes and common nomenclature and released the initial standards for grades of dressed beef that same year. Due to additional observation and experience, these standards continued to be modified and improved several times over the next few years.

In January 1923, the first meat-grading service was brought about as requested by the U.S. Shipping Board, as a result of their problems with purchasing consistently high-quality beef. The board asked that the Livestock, Meat, and Wool division of USDA grade beef carcasses based on the tentative U.S. standards previously published by USDA. In 1924, Congress authorized the federal grading of livestock and meat with the United States Agricultural Products Inspection and Grading Act. The help and cooperation of the National Live Stock and Meat Board with USDA was instrumental in making the arrangements for the application of federal grading and stamping services at all federally inspected meat plants.

Ten years after the release of the first grading standards, the grades were reworked and modified in order to include the suggestions and opinions of producers, packers, academic and industry professionals expressed at several hearings held in 1925 in an attempt to improve the grading standards. In 1926, the official U.S. Standards for Market Classes and Grades of Carcass Beef were published. These grades went into effect as a voluntary service for a one-year trial basis in plants the following year. After the one-year trial, the USDA meat-grading service was able to continue as a voluntary service. However, there was resistance and opposition from packers about the federal grading and stamping system. Between 1927 and 1938, several of the major packers developed their own grading and stamping systems. In fact, the Institute of American Meat Packers (IAMP) published their own set of grading standards in 1931, which encompassed 10 grades of beef. The plethora of grading systems at that time led to mass confusion between producers, retailers, and consumers alike; thus conferences between USDA and IAMP led to the exclusive use of the official U.S. standards.

The agency known to house the responsibility of grading and maintaining grading standards today, the Agricultural Marketing Service (AMS) of the USDA, gained these responsibilities from the Bureau of Agricultural Economics in 1939. Several significant modifications to the beef-grading standards have been made over the past 70 years. A single standard was adopted for steer, heifer, and cow beef in 1939, and Medium, Common, and Lower Cutter were revised to Commercial, Utility, and Canner, respectively. Grades for beef carcasses most common to those used today (Prime, Choice, Good, Standard, Commercial, Utility, Cutter, and Canner) were established in 1941. In addition, fat color was eliminated from the grading standards in 1949 with more changes to come. A year later, quality requirements were lowered by one full grade to shift what was then Choice into Prime, Good into Choice, and making two new divisions of Commercial. In 1956, Standard became the top half of Commercial when it was divided again based on maturity.

In 1962, a trial was conducted with the application of two types of grading systems: Quality and Cutability. As a result of this trial, the cutability standards, what were to be changed to "Yield Grades" in 1973, were adopted in 1965. In the same year, the grading standards were changed to lower the emphasis with regard to maturity for Prime, Choice, Good, and Standard grades as well as to make clear that all carcasses must be ribbed prior to grading. A class was created to designate those carcasses showing masculine or testosterone-induced characteristics associated with that of a young bull. It was known that these types of traits may have adverse affects on the ultimate eating experience of the retail cuts from this type of carcass; thus the "bullock" designation was created in 1973. In 1975, several grade changes were made, which molded the current standards in place today:

(1) For all "A" maturity steer, heifer, and cow beef, as well as bullock beef, maturity was eliminated for quality-grade determination.
(2) Marbling requirements necessary for the Good grade in "A" maturity were increased.
(3) In the Good and Standard grades, maximum maturity was lowered for steer, heifer, and cow beef in line with that allowed in Prime and Choice grades.
(4) Conformation scores were eliminated from quality-grade determination.

In addition, after 1975, carcasses had to be evaluated for both a quality and yield grade. Therefore, the 8 quality grades and 5 yield grades recognized amounted to 40 possible combinations.

In 1980, it was set forth that grading could only be performed on the carcass form; thus no wholesale cuts could be graded. The National Consumer Beef Retail Study (Savell et al., 1987) encouraged the evolution of the U.S. Good grade to U.S. Select in 1987 to better fit consumer attitudes and perceptions. In 1989, quality and yield grades were uncoupled so that carcasses did not have to be graded for both quality and yield in order for those carcasses that were trimmed on the slaughter floor to still be eligible for quality grading. The latest grade change came in 1997, when B-maturity carcasses with Small or Slight amounts of marbling were no longer eligible for U.S. Select and would now be recognized as U.S. Standard.

Marbling

Marbling (intramuscular fat) is the fat deposited between muscle fibers and bundles, believed by some to be associated with tenderness, juiciness, and flavor. For grading purposes, marbling is evaluated in the *M. longissimus thoracis* at the twelfth and thirteenth rib-interface. Subjective assessments performed by USDA graders determine what degree of marbling is present, which, in combination with the skeletal and lean maturity, determines the quality grade. As previously stated, as quality grade increases, overall palatability and its variability decrease as shown in Figure 6.1.

However, research both supports and opposes the theory of a relationship existing between marbling and palatability. A study comparing loin and round cuts from carcasses with a higher degree of marbling to those from a carcass with a lower degree of marbling found those from the former to be more juicy, flavorful, and tender, at least one-third of the time (Smith et al., 1984). In actuality, in loin steaks from an A-maturity carcass, marbling score has been shown to account for 24 to 34 percent of the variation in flavor, tenderness, and overall palatability. Another study did not observe the same relationship with respect to marbling and palatability (Brooks et al., 2000). These authors stated that when the degree of marbling observed is narrowed, this effect is substandard to that of a wide range of marbling

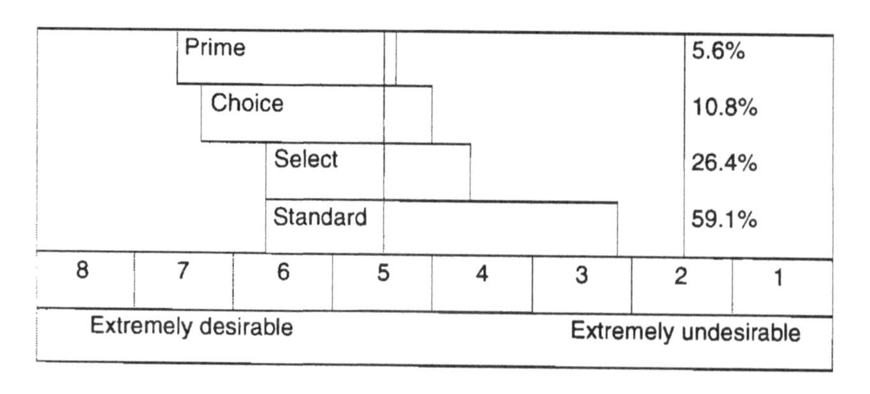

FIGURE 6.1. Percent of loin steaks receiving desirable and undesirable overall palatability ratings. *Source:* Adapted from information presented in Smith et al. (1987) and Miller (2004a).

scores. However, even when consumer sensory scores for "overall like" were not affected by marbling score, flavor scores were.

Marbling Theories

Four marbling theories are known for supporting the association of marbling with palatability: Lubrication, Insurance, Strain, and Bulk Density (Savell and Cross, 1988). The Lubrication Theory credits the marbling in and around muscle fibers for lubricating the mouth during the chewing process, thus making the meat juicier and more tender. The Insurance Theory states that with a higher degree of marbling, meat may be cooked to a higher degree of doneness and maintain the same eating quality. The Strain Theory suggests that when marbling is deposited into connective tissue walls, these walls are weakened and as a result the overall strength of connective tissue is weakened as it relates to tenderness. The Bulk Density Theory proposes that fat is softer and less dense than meat protein, which in turn, with fat present, makes a meat product more tender by decreasing its density. Thus, in a bite-size portion of meat, higher amounts of marbling decrease the mass per unit volume.

Window of Acceptability

Marbling is not only debated in terms of meat palatability, but is debated in dietary and health realms as well. With the increased intensity and focus on dietary fat in the 1980s, the optimal degrees of marbling for both eating experience and health-consciousness were assessed (Savell and Cross, 1988). Savell and Cross alluded that 3 to 7.3 percent was the "Window of Acceptability" in order for cuts to be healthy and to provide a pleasurable eating experience (Figure 6.2).

Muscle Specific

Research has shown that the association with quality grade (a correlated representation to the amount of intramuscular fat present) and palatability factors is much more pronounced in loin and ribs steaks as opposed to round steaks, where this relationship was present but at a much lesser extent (Smith et al., 1984). In fact, it has been suggested that overall maturity is more highly related to the eating expe-

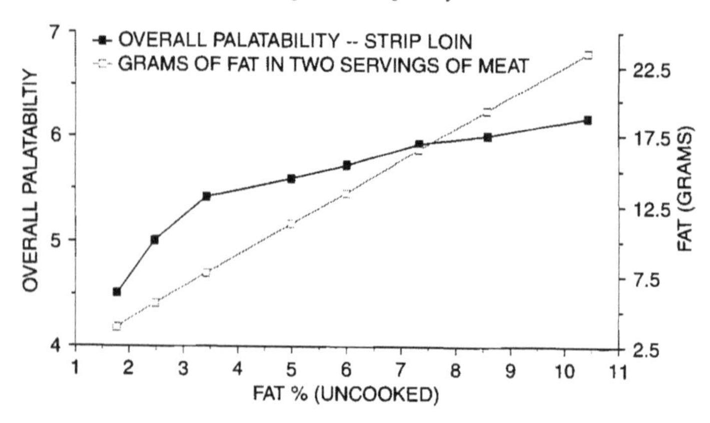

FIGURE 6.2. The role of fat in the palatability of beef, pork, and lamb. *Source:* Adapted from Savell and Cross (1988).

rience of round steaks than marbling. In addition, it has been shown across different primal cuts that as maturity increases, the role that marbling plays is decreased in regard to palatability. However, when comparing those muscle groups in more of a skeletal support role as opposed to those muscle groups functioning in a locomotive sense, there are various other factors that are known to be attributable to affecting palatability.

Maturity

Maturity is an evaluation of the physiological age of an animal and is primarily used as a predictor of carcass quality when combined with other factors (Aberle et al., 2001). As an animal matures and increases with age, various changes occur that affect the color, texture, and firmness of the lean.

Color

The change most visible and easily identified by the consumer is the change in color of lean. This change in color is due to the increase of myoglobin content in the muscle. Young cattle identified as veal (less than three months) contain 2 mg of myoglobin per gram of lean and a calf (three to nine months) will have 4 mg/g. These two age groups will exhibit a light, grayish pink color of lean. Young beef (one to two years) will increase to 8 mg/g and will have slightly dark red color, while cattle classified as old beef (greater than two years) will contain the highest amount, 18 mg/g, of myoglobin and will have lean color moderately dark red to dark red (Savell, 2004; Savell and Smith, 2000). The significance of this is due to the role that myoglobin plays in meat color. Myoglobin is the major source of pigment within the lean as it accounts for 80 to 90 percent of the total muscle pigment and is responsible for holding the O_2 that it obtains from hemoglobin (Miller, 2004b). Hemoglobin is the primary iron-containing protein pigment in blood and transports O_2 from the lungs to the muscle. As an animal increases in age, myoglobin loses its affinity to oxygen (Savell, 2004). As a result, there is an increase in myoglobin that is needed in order to maintain the transfer of O_2 to the muscle, causing the lean to become darker in color.

Texture and Firmness

In cattle, texture and firmness are additional characteristics of quality that are influenced by maturity. As an animal matures, the lean will progressively change in regard to texture and firmness, especially in steer, heifer, and cow beef. In the very youngest carcasses, the lean will possess a very fine texture and will be firm, allowing it to maintain its shape. As the animal matures the muscle will become progressively coarser and lose its firmness, resulting in very coarse, soft lean in mature beef (Savell and Smith, 2000).

Effect on Carcass Quality

Color, texture, and firmness are valuable in assessing carcass quality in beef. Together they can appropriately determine the level of maturity present within a carcass by assessing the physical changes expressed in the muscle as the animal matures. This influence on quality is especially important in consumer evaluation. Consumers tend to "shop with their eyes" and interpret the initial level of quality within a product by evaluating these attributes. As animals age, their lean becomes darker, losing its appeal to consumers. Smith et al. (1982) found that cuts from A-maturity carcasses are more likely to be assigned higher sensory panel scores in flavor, juiciness, tenderness, and overall palatability, as well as having low shear force values in comparison to cuts from carcasses of older maturity groups. On the other hand, beef from carcasses of C, D, and E maturity are less desirable in flavor, tenderness, and amount of connective tissue, in addition to overall palatability. However, these maturity groups have been shown to be more desirable in juiciness when compared to A- and B-maturity groups.

Furthermore, differences in palatability between maturity groups vary depending upon which maturity groups are compared. Small differences exist in tenderness of beef within youthful maturity groups (i.e., A versus B maturity), but substantial differences are found in tenderness when comparing lean from youthful and mature beef (i.e., A versus E maturity) (Smith et al., 1982).

Quality Grading Systems

The United States has implemented a hierarchical grading system, which is uniformly accepted throughout the nation and is practiced to ensure a common language and understanding among all participants of the industry.

U.S. Standards

As previously mentioned, quality grades are based upon the combination of an animal's marbling score and maturity. The major determinant of quality is the marbling score, which is determined by evaluating the amount of intramuscular fat within the exposed lean of the *M. longissimus thoracis* between the twelfth and thirteenth rib. Furthermore, maturity, which is determined by evaluating both lean and skeletal maturity, has been proven to influence the palatability of the final product. Skeletal maturity is assigned by evaluating the cartilaginous tip of the first three thoracic spinous processes, commonly referred to as "buttons," located beneath the split surface of the muscle. Physiological age is determined by examining the amount of cartilage that has been converted to bone, or percent of ossification, within the cartilaginous junctures. Lean maturity uses the color standards mentioned in the previous section to assign a lean maturity score. As demonstrated in Table 6.1, there are typically five classifications ranging from A to E, with the lean and maturity scores averaged together to obtain a final maturity score.

TABLE 6.1. USDA maturity group and approximate chronological age.

Carcass maturity	Approximate live age
A	9 to 30 months
B	30 to 42 months
C	42 to 72 months
D	72 to 96 months
E	More than 96 months

Source: USDA (1997).

Upon the determination of a final maturity score, an overall quality grade can be established based upon the combination of the maturity and marbling scores. The relationship between these two evaluations and how they influence quality grading is demonstrated in Figure 6.3. Prime is considered the highest grade as it possesses the greatest amount of marbling in a young maturity carcass. In contrast, Cutter is the least desirable as it provides the least amount of value and usefulness for the industry (Savell and Smith, 2000).

Other International Standards

Internationally, the practice of segregating cattle into various groups based upon their carcass evaluations is present in Canada, the European Union, Japan, and Australia, to name a few countries. The Canadian beef-grading system is designed to parallel the system implemented in the United States. The grade standards are set by the Federal Government in response to recommendations made by the Industry/Government Consultative Committee on Beef Grading. This system is voluntary and involves the use of quality grades indicating the level of expected palatability. In Canada, the quality grade is coupled with a yield grade designed to predict the percent of lean yield (Canadian Beef Grading Agency). Quality grades are labeled as Canada A, AA, AAA, and Prime, with the most common grades of Canada AA and AAA reported in 2003 (Table 6.2).

These grades are primarily based upon the degree of intramuscular fat present within the exposed muscle and the marbling score that is appropriately assigned, in addition to minor factors such as youthful-

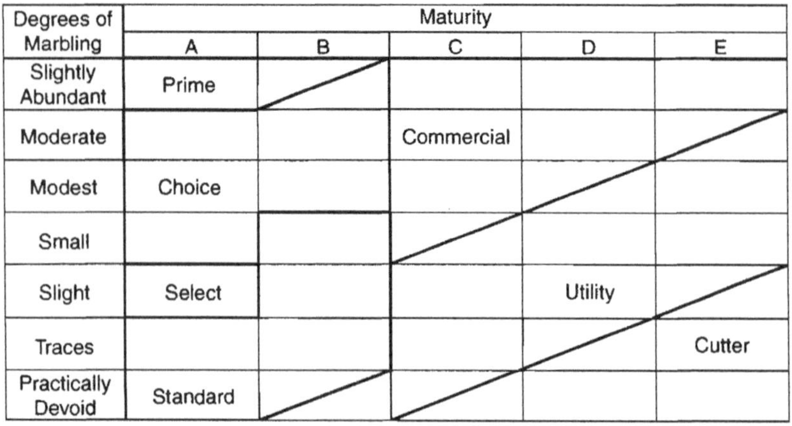

Degrees of Marbling	Maturity				
	A	B	C	D	E
Slightly Abundant	Prime				
Moderate			Commercial		
Modest	Choice				
Small					
Slight	Select			Utility	
Traces					Cutter
Practically Devoid	Standard				

FIGURE 6.3. Relationship between marbling, maturity, and carcass quality grade. *Source:* Adapted from Bovine Engineering, www. bovineengineering.com (2005).

TABLE 6.2. Canadian yield grades.

THE QUALITY GRADES						
Grade	**Maturity (age)**	**Muscling**	**Ribeye muscle**	**Marbling**	**Fat color and texture**	**Fat measure**
Canada Prime	Youthful	Good to excellent with some deficiencies	Firm, bright red	Slightly abundant	Firm, white or amber	2 mm or more
Canada A, AA, AAA	Youthful	Good to excellent with some deficiencies	Firm, bright red	A–trace AA–slight AAA–small	Firm, white or amber	2 mm or more

Source: Adapted from www.beefgradingagency.ca (2005).

ness of the carcass and color of lean within the *M. longissimus thoracis* at the twelfth and thirteenth rib interface (Canadian Beef Grading Agency, 2005; Saskatchewan Agriculture, Food and Rural Revitalization, 2000; McGrath, 1999). A carcass that has qualified for Canada Prime or any of the Canada A grades is then considered eligible to receive a yield grade.

The European Union (EU), Japan, and Australia have each adopted classification systems within their own respected beef industries. The EU segregates carcasses into groups based upon a variety of factors without promoting or enhancing the value of any one group or type of carcass. The EU executes the simplest evaluation system as it relies upon a conformation score and a fatness measurement as the primary indicator of classification in addition to the consideration of other variables such as weight, sex, and age. The conformation score consists of five main classes designated with the letters EUROP. These classes range from superior muscling (E) such as a double-muscled carcass, to a poorly muscled carcass (P) that would exhibit concave muscle structure and poor muscle development. The degree of fat consists of five classes ranging from a score of 1, with no fat to a low fat cover, and 5, where the entire carcass is covered with fat and possesses heavy fat deposits within the thoracic cavity. The most common type of beef carcass found in this region exhibits a conformation score of R and a fatness level of a low 4 (Council Regulation, 1981).

The Japanese Meat Grading Association combines yield and quality measurements to determine a classification score. Although it does strive to reach a certain type of carcass with ideal characteristics, it does not promote or reward those who produce carcasses with a specific grade. Quality is assessed by visual appraisal of the amount of marbling, color and brightness, firmness, and texture of the lean within the *M. longissimus thoracis,* coupled with the color, luster, and quality of the subcutaneous or outermost layer of fat exposed between the sixth and seventh rib (Japan Meat Grading Association, 1988).

The Australian grading system was defined by Meat Standards Australia (MSA), which began as an industry program in 1996 with the purpose of identifying factors associated with the decline of beef consumption within the country. The MSA grades were established in order to guarantee a product with a reliable level of eating quality (Meat and Livestock Australia Limited, 2003). This guarantee is based upon the evaluation of carcass factors and the establishment of acceptable standards. The factors that are included in this classification system consist mostly of quality factors, but also include other indicators in addition to the ones identified for the EU and Japan. Quality is assessed with a marbling score, which is defined by evaluating the amount, piece size, and distribution of the intramuscular fat at the ribbing site. The grade score is incrementally scaled (by 10 increments per unit) in range from 100 to 1100 based upon standards set by MSA (Meat and Livestock Australia Limited, 2003). Finally, overall maturity is determined by coupling the lean and skeletal maturity as discussed previously within the United States.

Additional factors used for classification include the evaluation of hump height, ultimate pH, color of both lean and fat, and the amount of subcutaneous fat visible over the carcass. Hump height is used to determine tropical breed content and is measured with a ruler parallel to the ribs including the dorsal edge of the ligamentum nuchae and across the dorsal surface of the *M. rhomboideus* (Meat and Livestock Australia Limited, 2003).

In addition to assigning a marbling score, the *M. longissimus thoracis* allows for the evaluation of pH and lean color. The pH is measured with a probe to determine the amount of lactic acid present within the *M. longissimus thoracis*. The color of lean is assessed on a chilled carcass and is scored against the Meat Color Standards determined by AUS-MEAT (Meat and Livestock Australia Limited, 2003).

Finally, an evaluation of fat is made by measuring the amount of subcutaneous fat in millimeters at a specified point within the rib section. The color of the fat on a chilled carcass is assessed at a point lateral to the *M. longissimus thoracis* and adjacent to the *M. iliocostalis*. This evaluation is determined by the AUS-MEAT Fat Color Reference Standards. Carcasses that fail to meet standards set by MSA are categorized as a non-MSA product (Meat and Livestock Australia Limited, 2003).

RELATIONSHIP BETWEEN CARCASS QUALITY AND CONSUMER ACCEPTABILITY

It has been difficult for scientists to find an accurate relationship between quality grade and consumer acceptability. Consumer acceptance of beef not only depends on quality grade, but also on the geographic location of the consumer, retail cut selected, cooking style used, and other parameters. Even though quality grades are not exact in their prediction of beef palatability, this system is the best indicator of quality segregation and consumer acceptance that is known today. Scientists are currently working on other technologies and taking other objective measurements of carcasses (i.e., hump height in Brahman carcasses) that, in the near future, might help reduce the variation within USDA quality grades.

As discussed in the previous section, overall maturity and marbling are the two factors used to determine quality grades. To better understand how consumer acceptance varies with quality grades, the variation of palatability within these two components must first be understood.

Overall Maturity

Maturity groups can help in the prediction of eating satisfaction. Smith et al. (1982) observed that the sensory panel assigned higher palatability ratings to broiled steaks that came from A maturity in comparison to B, C, and E carcasses in 62 to 86 percent of comparisons. Data found in this study suggest that maturity groups effectively segment carcasses that differ in flavor, tenderness, and overall palatability. However, maturity groups cannot exactly predict eating satisfaction; in fact, there is a decrease in the prediction of eating satisfaction as maturity increases (Smith et al., 1982).

Marbling

Some consumers think that marbling is a better indicator of beef palatability than maturity. Savell et al. (1987) suggested that consumer acceptance of beef depends not only on the amount of marbling but also on the geographic location of the consumer. The National Consumer Retail Beef Study on palatability evaluations of beef loin steaks that differed in marbling (Savell et al., 1987) was a consumer survey done in several cities of the United States. This study revealed that consumers from Philadelphia (known for being a Choice market) were more critical; they rated steaks with high marbling the same as consumers from Kansas City and San Francisco, but ratings were sharply reduced as marbling decreased from Slightly Abundant to Traces. On the other hand, Kansas City and San Francisco consumers gave fairly consistent high-palatability ratings to the same steaks and slightly reduced ratings as marbling decreased. Savell et al. (1987) concluded that consumer acceptance of steaks with varying marbling is not consistent throughout the United States and these differences need to be accounted for when marketing beef.

A few years prior, Smith et al. (1984) had found that loin and round steaks that came from A-maturity carcasses with "Moderately Abundant" marbling were more acceptable than the same steaks that had "Practically Devoid" marbling. However, acceptability of round steaks did not decrease with marbling scores in a constant manner as loin steaks did.

Quality Grade

Several studies have been conducted in the United States addressing quality grades and palatability of beef (Smith et al., 1987; Neely et al., 1998). The National Consumer Retail Beef Study (Savell et al., 1989), this time evaluating the interaction of trim level, price, and quality grade on consumer acceptance of beef steaks and roasts, found that consumers would purchase a particular quality grade depending upon the city and not on price. For instance, San Francisco consumers preferred to purchase Select steaks instead of Choice steaks lower in price. They also observed that neither consumers from San Francisco nor Philadelphia could observe visual differences between Choice and Select grades.

One of the most important findings of the National Consumer Retail Beef Study (Savell et al., 1989) was that consumers highly accepted U.S. Choice and Select carcasses but for different reasons. U.S. Choice was accepted for taste but consumers criticized the amount of fatness and U.S. Select was accepted for leanness but was criticized for the taste (flavor) and texture (tenderness and juiciness) of steaks. This implies that each grade can be marketed for its unique attributes to satisfy the consumer's personal preference (Savell et al., 1989).

Other data show that more than 80 percent of the time USDA Commercial, Standard, Select, Choice, and Prime will be acceptable (Smith et al., 1987). Also, the three highest grades (Prime, Choice, and Good [now Select]) almost assured (≥90 percent) acceptability of broiled loin steaks in all palatability attributes and of round steaks in juiciness, tenderness, and amount of connective tissue (Smith et al., 1987).

Quality grade is not the best indicator of consumer acceptance (Carpenter, 1974 as cited by Smith et al., 1987) but it is the best indicator that is available for the meat industry currently. Consumers mainly accept Prime, Choice, and Select quality grades, and consumer acceptability drastically reduces when they are presented with Commercial, Utility and Cutter grades of loin steaks (Smith et al., 1987). Neely et al. (1998) state that "customer satisfaction for beef steaks is a complex issue because of the interrelated effects of cut, USDA quality grade, and city on palatability. Targeted marketing of beef to consumers using a well-defined plan may be the logical system needed to ensure maximum eating enjoyment" (p. 1033).

RELATIONSHIP BETWEEN MOUTHING AND SKELETAL MATURITY

Skeletal Maturity

As previously mentioned, maturity can be determined by the evaluation of size, shape, and ossification of the bones and cartilage, especially the split chine bones, and the color and texture of the lean (USDA, 1997). In the U.S. Standards for Grades of Carcass Beef (USDA, 1997), the percentage of ossification in the thoracic buttons of the split chine, ossification of the sacral vertebra, and rib color, size, and shape are evaluated to determine skeletal maturity. Also in the Standards for Grades (USDA, 1997) lean maturity factors assessed are the color, texture, and firmness of the *M. longissimus thoracis.*

Dentition

Currently in the U.S. beef industry, maturity has become an increasingly prevalent issue because of the recent BSE (Bovine Spongiform Encephalopathy) outbreaks. This has caused evaluation of alternate aging-determination processes. Using dentition to age cattle has become the most recently accepted practice. According to the Food Safety and Inspection Service (FSIS) within the USDA (FSIS, 2004), examination of teeth serves as the best and most practical method of age determination.

Cattle are born with deciduous teeth, commonly referred to as milk teeth. These teeth are lost as the animal ages and are replaced with permanent teeth. Deciduous teeth eruption somewhat varies but most calves have all teeth erupted at birth and permanent teeth replace deciduous teeth as the animal ages. Deciduous teeth appear to be much smaller and narrower than permanent teeth (FSIS, 2004).

Three types of teeth are found in bovine: incisors, premolars, and molars. Incisors are absent from the upper jaw and are in the rostral portion of the mouth. Commonly known as cheek teeth, molars and premolars are present in both the caudal part of the upper and lower jaws. Figure 6.4 illustrates a mature bovine skull with all permanent teeth present and Table 6.3 illustrates the eruption times of permanent teeth.

Eruption of teeth in cattle typically follows the pattern shown in Figure 6.5. In Figure 6.5-A, all deciduous teeth are present and becoming looser in Figure 6.5-B. Figure 6.5-C illustrates one permanent incisor (I 1) has erupted. In Figure 6.5-D and 6.5-E, both central incisors (I 1) have straightened and the inside corners are in line. Figure 6.5-F illustrates when both middle incisors (I 2) have erupted and the animal is thought to have been between 24 and 30 months of age.

Relationship Between Dentition and Skeletal Maturity

Dentition has been used to evaluate cattle age for many years; however, limited data are available on the relationship between chro-

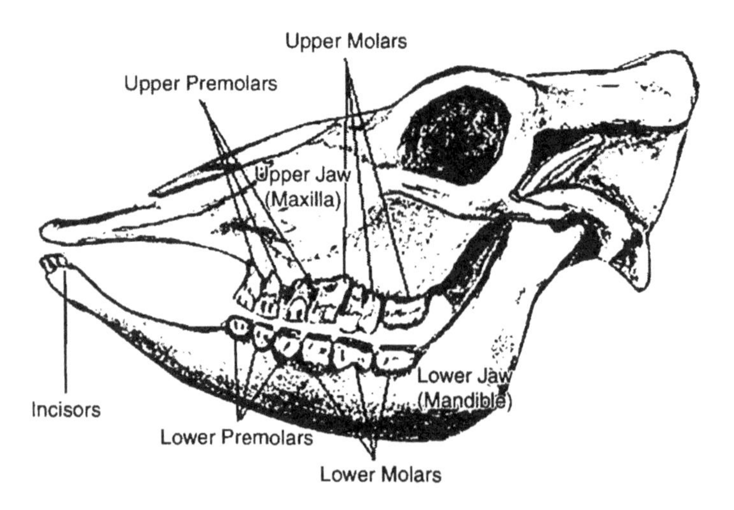

FIGURE 6.4. Mature bovine skull with all permanent teeth present. *Source:* FSIS (2004).

TABLE 6.3. Eruption times of permanent teeth.

Teeth	Age at eruption
First Incisor (I 1)	18 to 24 months
Second Incisor (I 2)	24 to 30 months
Third Incisor (I 3)	36 months
Fourth Incisor (I 4 or C)	42 to 48 months
First Cheek Tooth (P 2)	24 to 30 months
Second Cheek Tooth (P 3)	18 to 30 months
Third Cheek Tooth (P 4)	30 to 36 months
Fifth Cheek Tooth (M 2)	12 to 18 months
Sixth Cheek Tooth (M 3)	24 to 30 months

Source: FSIS (2004).

Note: I = Incisor; P = Premolar; M = Molar.

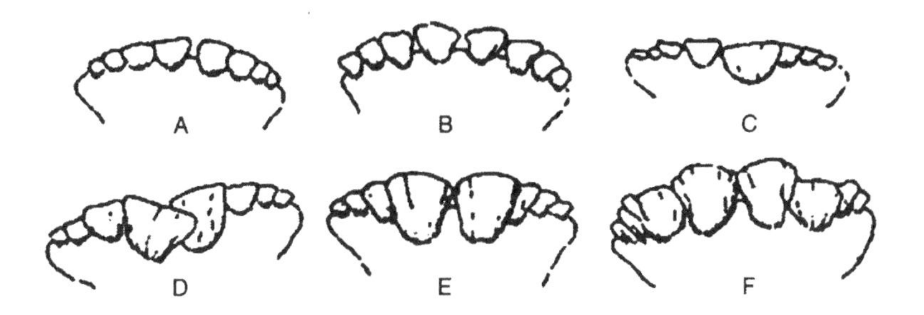

FIGURE 6.5. Progression of eruption of permanent incisors. *Source:* FSIS (2004).

nological age and USDA maturity score. Shackelford, Koohmaraie, and Wheeler (1995) found that carcass maturity was moderately related to chronological age ($r^2 = 0.60$) and that carcass maturity scores increased with increased chronological age at a much faster rate than indicated by USDA. Extrinsic factors have been shown to influence the skeletal maturity of the animal, including implant effects (Unruh, Gray, and Dikeman, 1986; Foutz et al., 1997 as cited by Lawrence, Whatley, Montgomery, and Perino, 2001) and spray chilling (Allen et al., 1987 as cited by Lawrence, Whatley, Montgomery, and Perino, 2001).

There also have been factors identified that affect the rate of eruption, therefore disturbing the rate of eruption of permanent incisors. Malnutrition has been shown to slow the eruption process (Brookes and Hodges, 1979; England, 1984; and Wass et al., 1986 as cited by Lawrence, Whatley, Montgomery, and Perino, 2001). Gender also has shown to affect the rate of eruption, with heifers typically erupting permanent incisors later than steers and bulls (Andrews and Wedderburn, 1977). Estrogen concentration in the animal also can be a factor influencing skeletal maturity and was illustrated by Silberberg and Silberberg (1939), who concluded that estrogen administered to immature guinea pigs caused premature ossification of cartilage in the epiphysial disks, ribs, and vertebra. Andrews and Wedderburn (1977) also demonstrated that there appears to be a breed difference in eruption of permanent incisors. Andrews (1975) reported that aging cattle by teeth can be extremely variable depending on gender, breed, and nutrition during feeding.

Lawrence, Whatley, Montgomery, and Perino (2001) suggested that dentition is a more accurate and objective measurement of carcass maturity; however, they found no evidence to support dentition being a better predictor of lean palatability. Lawrence, Whatley, Montgomery, and Dikeman (2001) found no differences in Warner-Bratzler shear force, or sensory panel evaluations for tenderness among the five dental classes.

Countries other than the United States have already implemented dentition in their classification systems; for instance, Australia and South Africa. Both countries use the number of incisors present at slaughter to estimate maturity (Government Gazette, 1990; AUS-MEAT, 1995).

SUMMARY

Beef quality encompasses many different sensory properties, but is primarily defined by tenderness, juiciness, and flavor. Grading, designed to sort carcasses based on specific traits, is our best estimate of beef quality in the United States. Marbling and maturity are the two primary factors involved in grading beef carcasses. Grading systems vary internationally due to consumer preferences in specific regions. Consumer acceptance of beef is very difficult to define universally, as it is not only dependent on quality grade, but also on the geographic location of the consumer, retail cut selected, cooking style used, and other parameters. Maturity has also become an increasingly prevalent issue in the United States because of the recent BSE incidences. Dentition has been identified by FSIS as the best way to age cattle and has become the most recently accepted practice. Much effort has been and will continue to be dedicated to improving quality in the beef industry in the future.

LITERATURE CITED

Aberle, E.D., J.C. Forrest, D.E. Gerrard, and E.W. Mills. 2001. *Principles of Meat Science,* 4th ed. Kendall/Hunt Publishing Co., Dubuque, IA.

Allen, D.M., M.C. Hunt, A.L. Filho, R.J. Danler, and S.J. Goll. 1987. Effects of spray chilling and carcass spacing on beef carcass cooler shrink and grade factors. *J. Anim. Sci.* 64: 165-170.

Andrews, A.H. 1975. The relationship between age and development of the anterior teeth in cattle as determined by the oral examination of 2900 animals between the ages of 12 and 60 months. *Br. Vet. J.* 131: 152-159.

Andrews, A.H. and R.W.M. Wedderburn. 1977. Breed and sex differences in the age of appearance of the bovine central incisor teeth. *Br. Vet. J.* 133: 543-547.

AUS-MEAT. 1995. *Handbook of Australian Meat,* 5th ed. Australian meat and livestock corporation, Brisbane, Queensland, Australia.

Bovine Engineering. 2005. USDA quality grades. Available at: www.bovine engineering.com. Accessed July 29, 2005.

Briskey, E.J. and R.W. Bray. 1964. A special study of the beef grade standards, mimeograph report submitted to the American National Cattleman's Association, Denver, CO.

Brookes, A.J. and J. Hodges. 1979. Breed, nutritional and heterotic effects on age of teeth emergence in cattle. *J. Agric. Sci.* 93: 681-685.

Brooks, J.C., J.B. Belew, D.B. Griffin, B.L. Gwartney, D.S. Hale, W.R. Henning, D.D. Johnson, J.B. Morgan, F.C. Parrish Jr., J.O. Reagan, and J.W. Savell. 2000. National beef tenderness survey—1988. *J. Anim. Sci.* 78: 1852-1860.

Canadian Beef Grading Agency. Available at: www.beefgradingagency.ca. Accessed July 2, 2005.

Carpenter, Z.L. 1974. Beef quality grade standards—need for modifications? Proc. Recip. Meat Conf. 27: 122-142.

Council Regulation (EEC) No. 1208/81. 1981. Determining the community scale for the classification of carcasses of adult bovine animals. *OJ L* 123, 7.5: 3.

England, R.B. 1984. Incisor tooth eruption, development and attrition. *Agri-Practice* 5: 44-52.

Foutz, C.P., H.G. Dolezal, T.L. Gardner, D.R. Gill, J.L. Hensley, and J.B. Morgan. 1997. Anabolic implant effects on steer performance, carcass traits, subprimal yields, and longissimus muscle properties. *J. Anim. Sci.* 75: 1256-1265.

FSIS. 2004. BSE information: Using dentition to age cattle. Food Safety and Inspection Service, USDA, Washington, DC. Available at: www.fsis.usda.gov/OFO/TSC/bse_information.htm. Accessed March 16, 2005.

Government Gazette. 1990. South African agricultural product standards act (No. 119). Regulations regarding the classification marketing of meat. Government Gazette, Pretoria, South Africa.

Harris, J.J., H.R. Cross, and J.W. Savell. 1990. History of meat grading in the United States. Available at: savell-j.tamu.edu/history.html. Accessed May 23, 2002.

Japan Meat Grading Association. 1988. New beef carcass grading standards. Tokyo, Japan.

Lawrence, T.E., J.D. Whatley, T.H. Montgomery, and L.J. Perino. 2001. A comparison of the USDA ossification-based maturity system to a system based on dentition. *J. Anim. Sci.* 79: 1683-1690.

Lawrence, T.E., J.D. Whatley, T.H. Montgomery, L.J. Perino, and M.E. Dikeman. 2001. Influence of dental carcass maturity classification on carcass traits and tenderness of longissimus steaks from commercially fed cattle. *J. Anim. Sci.* 79: 2092-2096.

McGrath, S. 1999. Ultrasound, carcass value and genetic evaluation. Canadian Limousin Association. Available at: www.limousin.com/pages/breed_improvement/article_8.htm. Accessed August 5, 2004.

Meat and Livestock Australia Limited. 2003. Meat Standards Australia Carcass Grading System. ABN 39 081 678 364. Locked Bag 991 N. Sydney NSW 2059, Australia.

Miller, R.K. 2004a. Animal Science 647. Class notes. Department of Animal Science, Texas A&M University, College Station, TX.

Miller, R.K. 2004b. Meat color. Class notes. *Anim. Sci.* 647. Lecture 11. Texas A&M University, College Station, TX.

Neely, T.R., C.L. Lorenzen, R.K. Miller, J.D. Tatum, J.W. Wise, J.F. Taylor, M.J. Buyck, J.O. Reagan, and J.W. Savell. 1998. Beef customer satisfaction: Role of cut, USDA quality grade, and city on in-home consumer ratings. *J. Anim. Sci.* 76: 1027-1033.

Saskatchewan Agriculture, Food and Rural Revitalization. 2000. Live cattle ultrasound and the Canadian beef grading system. Available at: www.agr.gov.sk.ca/DOCS/livestock/beef/marketing/ultrasound.asp. Accessed July 2, 2005.

Savell, J.W. 2004. Animal Science 307. Class notes. Department of Animal Science, Texas A&M University, College Station, TX.

Savell, J.W., R.E. Branson, H.R. Cross, D.M. Stiffler, J.W. Wise, D.B. Griffin, and G.C. Smith. 1987. National consumer retail beef study: Palatability evaluations of beef loin steaks that differed in marbling. *J. Food Sci.* 52 (3): 517-519, 532.

Savell, J.W. and H.R. Cross. 1988. The role of fat in the palatability of beef, pork, and lamb. In: *Designing Foods*. National Academy Press, Washington, DC, pp. 345-355.

Savell, J.W., H.R. Cross, J.J. Francis, J.W. Wise, D.S. Hale, D.L. Wilkes, and G.C. Smith. 1989. National consumer retail beef study: Interaction of trim level, price, and grade on consumer acceptance of beef steaks and roasts. *J. Food Quality* 12: 251-274.

Savell, J.W. and G.C. Smith. 2000. *Laboratory Manual for Meat Science,* 7th ed. American Press, Boston, MA.

Shackelford, S.D., M. Koohmaraie, and T.L. Wheeler. 1995. Effects of slaughter age on meat tenderness and USDA carcass maturity scores of beef females. *J. Anim. Sci.* 73: 3304-3309.

Silberberg, M. and R. Silberberg. 1939. Action of estrogen on skeletal tissues of immature guinea pigs. *Arch. Pathol.* 28: 340-360.

Smith, G.C., Z.L. Carpenter, H.R. Cross, C.E. Murphey, H.C. Abraham, J.W. Savell, G.W. Davis, B.W. Berry, and F.C. Parrish Jr. 1984. Relationship of USDA marbling groups to palatability of cooked beef. *J. Food Quality.* 7: 289-308.

Smith, G.C., H.R. Cross, Z.L. Carpenter, C.E. Murphey, J.W. Savell, H.C. Abraham, and G.W. Davis. 1982. Relationship of USDA maturity groups to palatability of cooked beef. *J. Food Sci.* 47 (4): 1100-1107, 1118.

Smith, G.C., J.W. Savell, H.R. Cross, Z.L. Carpenter, C.E. Murphey, G.W. Davis, H.C. Abraham, F.C. Parrish Jr., and B.W. Berry. 1987. Relationship of USDA quality grades to palatability of cooked beef. *J. Food Quality.* 10: 269-286.

Unruh, J.A., D.G. Gray, and M.E. Dikeman. 1986. Implanting young bulls with zeranol from birth to four slaughter ages: I. Live measurements, behavior, masculinity, and carcass characteristics. *J. Anim. Sci.* 62: 279-289.

USDA. 1997. Official United States Standards for Grades of Carcass Beef. Livestock and Seed Division. Agricultural Marketing Service, USDA, Washington, DC.

Wass, W.M., J.R. Thompson, E.W. Moss, J.P. Kunesh, P.G. Eness, and L.S. Thompson. 1986. Examination of the teeth and diagnosis and treatment of dental diseases. In: *Current Veterinary Therapy: Food Animal Practice 2,* J. Howard (ed). W.B. Saunders Company, Philadelphia, PA, pp. 713-714.

Chapter 7

Sensory Attributes and Quality

Rhonda Miller

INTRODUCTION

Sensory attributes of foods are those factors that affect the senses of sight, smell, taste, feel, or sound. For beef, the major sensory properties have been defined as color, juiciness, tenderness, and flavor. These four sensory attributes are measured through the human senses of sight, smell, taste, and feel. Therefore, consumers or humans use their senses to determine the quality of beef. They first see beef either in the package or on their plate prior to consumption. The visual appearance of the beef, either in the raw state or the cooked state, will influence perceptions of acceptability or preference. When beef is cut and eaten through mastication in the mouth, the senses of smell, taste, and feeling are used to determine the acceptability of the eating experience. The three sensory attributes of juiciness, tenderness, and flavor are often collectively referred to as beef palatability. While color is an extremely important sensory attribute, it is beyond the scope of this chapter and will not be further discussed.

The overriding assumption is that as beef becomes juicier, more tender, and more flavorful, beef is more palatable and more acceptable to consumers. While this relationship is not as simple as stated, in general, it is true. Meat scientists have used sensory attributes or meat palatability since the mid-1900s to describe pre- and postharvest factors that affect beef-eating quality. The sensory properties of beef

Handbook of Beef Safety and Quality

doi:10.1300/5640_07

are linked to beef-quality perceptions, even though beef quality is a broader term and sensory properties are one aspect of beef quality. Meat scientists have used sensory evaluation to understand how to improve the quality of beef and to help provide consumers with higher quality beef. In 1978, the American Meat Science Association published Sensory and Cookery Guidelines for Meat (AMSA, 1978). This document, updated in 1995 (AMSA, 1995), has become the guideline for conducting beef-sensory evaluation.

Sensory evaluation can be conducted using trained sensory panelists or consumer sensory panelists. The type of panelists, trained or consumer panelists, affects the methods, ballot, and interpretation of the sensory test. Each method has validity as a sensory tool and provides information concerning the relationship between beef sensory properties and quality.

The objective is to discuss sensory evaluation tools used to evaluate beef quality, define the major sensory properties of beef, provide an explanation of how each sensory property is related to beef quality, discuss how pre- and postharvest factors influence the sensory properties of beef, and how muscle location within a carcass influences sensory properties.

METHODS USED IN BEEF SENSORY EVALUATION

Trained Meat Descriptive Attribute Sensory Evaluation

The predominant method of sensory evaluation of beef quality has utilized trained sensory panelists and followed the procedures of AMSA (1978, 1995). This method uses Meat Descriptive Sensory Evaluation of juiciness, tenderness, and flavor using eight-point, verbally anchored scales. The method is descriptive as specific attributes are defined and the level of the attribute is quantified using trained sensory panelists. This provides a method to compare one product to another for intensity of specific attributes, juiciness, tenderness, and flavor attributes. Beef tenderness is many times divided into three attributes: muscle-fiber tenderness, connective tissue amount, and overall tenderness. A ballot using the Meat Descriptive Attribute Sensory evaluation for beef palatability is presented in Exhibit 7.1. This ballot

EXHIBIT 7.1. Example Meat Descriptive Attribute Sensory Ballot

Sample	Juiciness	Muscle-Fiber Tenderness	Connective Tissue Amount	Overall Flavor Intensity	Off-Flavor Characteristics

	Rating Scale	Rating Scale	Rating Scale	Rating Scale	Rating Scale
	8 Extremely Juicy	8 Extremely Tender	8 None	8 Extremely Intense	A Acid
	7 Very Juicy	7 Very Tender	7 Practically None	7 Very Intense	B Bitter
	6 Moderately Juicy	6 Moderately Tender	6 Traces	6 Moderately Intense	BR Browned
	5 Slightly Juicy	5 Slightly Tender	5 Slight	5 Slightly Intense	C Cardboard
	4 Slightly Dry	4 Slightly Tough	4 Moderate	4 Slightly Bland	CH Chemical
	3 Moderately Dry	3 Moderately Tough	3 Slightly Abdt	3 Moderately Bland	CW Cowy
	2 Very Dry	2 Very Tough	2 Slightly Abdt	2 Very Bland	F Fishlike
	1 Extremely Dry	1 Extremely Tough	1 Abundant	1 Extremely Bland	L Liver
					M Metallic
					N Nutty
					P Putrid
					SA Salty
					SB Serumy/bloody
					SD Soured
					SO Sour
					SW Sweet

Note: 8-point rating scales adapted from AMSA (1995).

shows the use of eight-point descriptive scales for beef palatability attributes of juiciness, the three attributes of muscle fiber tenderness, connective tissue amount, and overall tenderness, and flavor intensity. A column for panelists to identify off-flavors that may be detected during evaluation can also be used. Other attributes of interest can be added to the ballot and similar eight-point scales added on a product-specific basis.

Flavor and Texture Descriptive Attribute Evaluation

Other descriptive sensory methods have been applied to beef in the past 20 years. The Spectrum Method or the Quantitative Descriptive Analysis (QDA) sensory methods have been used to examine more specific flavor and texture attributes of beef. In these methods, more detailed definitions of flavor or texture are used and each attribute is quantified.

The Spectrum Method was developed by Gail Vance Civille and flavor and/or texture attributes for a product are defined. Panelists are extensively trained to identify and quantify each attribute. Panelists test the samples independently for each attribute and any other attributes that they identify. This method uses a dictionary of product descriptors called a lexicon. Johnsen and Civille (1986) defined a group of flavor attributes for beef flavor (Table 7.1). In this lexicon, specific flavors that are anchored with a definition can be measured in beef. Beef flavor is very complex and is made up of many different attributes. In the Spectrum Method, instead of evaluating the overall flavor in beef, the specific attributes of cooked beefy/brothy, cooked beef fat, cardboardy, cowy, grainy, serumy, painty, and livery may be measured. In addition, sensory attributes such as basic tastes of sweet, sour, bitter, and salty, mouthfeels such as astringency and metallic, after-tastes and -mouthfeels, and meat texture attributes such as springiness, juiciness, muscle-like bite, and hardness can be evaluated. To train a panel to use this lexicon, panelists are presented with a reference sample for each attribute; they are trained to identify the attribute at varying levels. For example, livery flavor is a common component of beef flavor. To train panelists to understand and identify livery flavor, cooked beef liver is presented to the panelists. This provides the panelist with the flavor attribute in a very concentrated form. To

TABLE 7.1. Lexicon for warmed-over flavors in beef.

Descriptors	Definition of descriptors
Aromatics	
Cooked beef lean	The aromatic associated with cooked beef muscle meat
Cooked beef fat	The aromatic associated with cooked beef fat
Browned	The aromatic associated with the outside of grilled or broiled beef
Serum/bloody	The aromatic associated with raw beef lean
Grainy/cowy	The aromatic associated with cow meat and/or beef with grain/feed character
Cardboard	The aromatic associated with slightly stale beef, refrigerated for a few days only and associated with wet cardboard and stale oils and fats
Painty	The aromatic associated with rancid oil and fat
Fishy	The aromatic associated with some rancid fats and oils
Livery/organy	The aromatic associated with beef liver and/or kidney
Basic Tastes	
Sweet	Taste on the tongue associated with sugars
Sour	Taste on the tongue associated with acids
Salty	Taste on the tongue associated with sodium ions
Bitter	Taste on the tongue associated with bitter agents such as caffeine, quinine, etc.

Source: Adapted from Johnsen and Civille (1986).

help panelists identify livery flavor when it is present with other flavor attributes, liver can be added to ground beef from very small to higher concentrations. Panelists will be asked to quantify the level of liver in the ground beef patties. This exercise helps panelists to understand how to identify liver in the presence of other attributes and how to identify the attribute at low to high levels. After the panelists understand each attribute, after they are anchored on each attribute and they can recognize and scale from low to high levels consistently for all attributes, the panel is considered trained and testing can begin. The panel leader oversees these exercises and consistently provides feedback to panelists so that, as a group, the panelists are rating attributes similarly. Panelists are tested for their ability to consistently rank sample attributes using the AMSA (1995) panel verification test. This test is conducted by presenting panelists with nine samples that vary in attributes. The nine samples are presented four times, once per day for four days. Within a day, sample order of presentation is randomized. The test evaluates the panelist's ability to consistently rank a sample for specific attributes over the four days. If panelists pass this test, they are allowed to evaluate samples for testing. During testing, panelists are presented with beef samples and they identify and quantify all flavor and/or texture attributes that they detect. Changes in the levels of all or any of these attributes can then be examined and a fuller understanding of changes in oral senses between products during consumption can be determined. An example of a ballot used in the Spectrum Method is presented in Exhibit 7.2.

The QDA sensory method is slightly different from the Spectrum Method and was developed by the Tragon Corporation as a method to determine differences in sensory attributes of food products. In this method, trained panelists are selected from a previously trained panelist pool. Panelists are selected based on their ability to identify flavor and/or texture attributes for the product of interest. Panelists are trained similarly as described for the Spectrum Method where the product and ingredient references are used to generate descriptive terms; however, the panel leader acts as a facilitator, not an instructor. Also, panelists develop their own scaling and vocabulary. A 15-cm line scale is used to quantify each sensory attribute. During testing, panelists independently evaluate each product. Panelists do not discuss data, terminology, or samples after each taste session. Feedback on performance relative to other panel members or differences between samples is provided to panelists through the panel leader. Data are entered into a computer, are analyzed statistically, and reported in graphic representation in the form of a "spiderweb" where a branch from the center point in the spider web represents the intensity of that attribute. This analysis takes into account differences between panelists and differences in terminology. An example of QDA results is presented in Figure 7.1. Differences in the spider web between products determine whether sensory differences exist. It is obvious from looking at this figure, that as marbling or chemical lipid increases in beef steaks, overall aroma, mouthcoating, and fat flavor increase and livery flavor decreases. In addition, beef steaks with 2 and 4 percent chemical lipid have similar beefy flavor; whereas beef steaks with 4 and 6 percent chemical lipid have similar serum/bloody/rare beef

EXHIBIT 7.2. Example Trained Sensory Panel Flavor and Texture Descriptive Attribute Sensory Ballot Using Spectrum Method Defined by Meilgaard et al. (1999)

Sample ID#	W/U	818	744	409
Aromatics:				
Cooked Beef/Brothy	____	____	____	____
Cooked Beef Fat	____	____	____	____
Serumy/Bloody	____	____	____	____
Grainy/Cowy	____	____	____	____
Cardboard	____	____	____	____
Painty	____	____	____	____
Fishy	____	____	____	____
Liver	____	____	____	____
Soured	____	____	____	____
Browned/Burnt	____	____	____	____
Other (describe)	____	____	____	____
Feeling Factors:				
Metallic	____	____	____	____
Astringent	____	____	____	____
Basic Tastes:				
Salt	____	____	____	____
Sour	____	____	____	____
Bitter	____	____	____	____
Sweet	____	____	____	____
Aftertastes:				
Astringent	____	____	____	____
Fat Mouth Feel	____	____	____	____
Bitter	____	____	____	____
Browned/Burnt	____	____	____	____
Sour	____	____	____	____
Sweet	____	____	____	____
Other (describe)	____	____	____	____
After-Feeling Factors:				
Lip Burn	____	____	____	____
Metallic	____	____	____	____
Texture:	____	____	____	____
Springiness	____	____	____	____
Hardness	____	____	____	____

Note: 0 = none and 15 = extremely intense for each attribute.

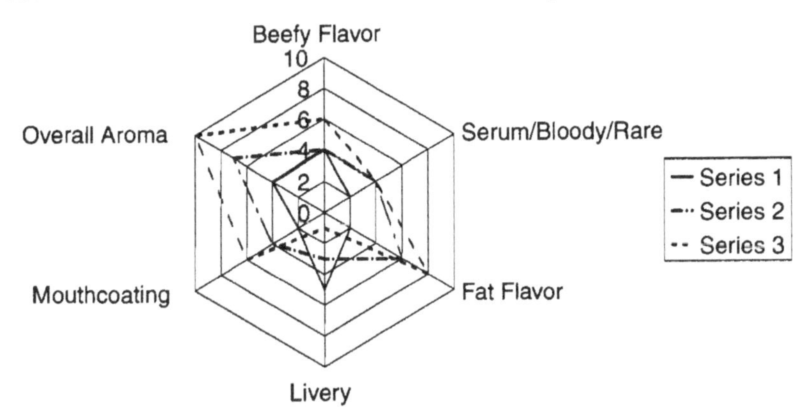

FIGURE 7.1. A spiderweb generated from sensory data using the Tragon Quantitative Descriptive Analysis Method for beef steaks varying in marbling level where Series 1 = 2 percent chemical lipid; Series 2 = 4 percent chemical lipid; and Series 3 = 6 percent chemical lipid.

flavor. These data are easy to interpret and it is easy to understand differences between steaks.

Consumer sensory evaluation also is an important sensory tool used to evaluate beef. Consumer sensory evaluation asks consumers questions concerning overall like or dislike, like or dislike for a specific attribute of the product (usually juiciness, tenderness, and flavor) or their preference for one product over another. It is very important when conducting consumer tests to select an unbiased, random group of consumers. Consumers are not trained. Consumer tests can be conducted in a central location where consumers come to a location and the testing environment is standardized, defined as Central Location Tests, and controlled like a sensory laboratory or testing site or they can be conducted in the consumer's home, called In-Home Consumer Tests. Central Location Consumer Tests are very controlled tests and do not include the family's opinion, and the environment is less real; whereas In-Home Consumer Tests include the family's opinion, but there is no control of outside factors that could impact consumer sensory verdicts. Consumer and trained sensory evaluations are separate sensory methods for determining beef quality, but they both are valid tools.

BEEF SENSORY ATTRIBUTES

Juiciness

Juiciness is defined as the amount of expressible juice perceived in the oral cavity during mastication. For beef, during chewing, moisture and fat from the product are released and perceived within the oral cavity. Juiciness can be evaluated as a single attribute or it is sometimes evaluated in two components, initial juiciness and sustained juiciness. Initial juiciness is the amount of moisture perceived within the first one to five chews and sustained juiciness is the amount of moisture perceived in the product after five chews and throughout the chewing processes prior to swallowing or expectorating the product. The length of time or the number of chews used for initial juiciness can be varied and should be defined for the product. Juiciness is usually rated either from none to extremely juicy or from extremely dry to extremely juicy. Meilgaard, Civille, and Carr (1999) defined juiciness as the amount of juice/moisture perceived in the mouth and can be measured using the Spectrum Scale where 0 = none and 15 = very juicy. While there are multiple scales for measuring juiciness, most of the scales have the comparable property of measuring the amount of expressible juices in the product during chewing. For example, Meilgaard, Civille, and Carr (1999) used a half-inch slice of banana as a 1 for juiciness and half-inch slice of a raw mushroom as a 4 for juiciness, a half-inch wedge of a red delicious apple as a 10 for juiciness and a half-inch cube of watermelon for a 15 for juiciness. Using the AMSA (1995) descriptive scale for juiciness in Exhibit 7.1, a 1 would be similar to placing a cotton ball in your mouth and an 8 would be similar to placing a half-inch cube of canned peach in your mouth. Juiciness is an attribute that is easy to understand and to identify by a trained panelist. Consumers also easily understand what is meant by juiciness of a product. For consumer sensory evaluation, consumers can be asked whether they like or dislike the juiciness of a product or they can be asked to rate the juiciness of a product using an end-anchored scale as shown in Exhibit 7.3. In Exhibit 7.3, Questions 8 and 9 of the ballot are examples of questions that consumers can be asked about their like/dislike of the juiciness of a product and the level of juiciness.

EXHIBIT 7.3. Example consumer sensory ballot for evaluation of beef steaks.

1. Indicate by placing a mark in the box your **OVERALL LIKE/DISLIKE** of this sample.

☐ ☐ ☐ ☐ ☐ ☐ ☐ ☐ ☐

Like extremely Neither Like nor Dislike Dislike extremely

2. Indicate by placing a mark in the box your **OVERALL LIKE/DISLIKE** for the **FLAVOR** of this sample.

☐ ☐ ☐ ☐ ☐ ☐ ☐ ☐ ☐

Like extremely Neither Like nor Dislike Dislike extremely

3. Indicate by placing a mark in the box how you feel about the **INTENSITY OF THE FLAVOR** of this sample.

☐ ☐ ☐ ☐ ☐ ☐ ☐ ☐ ☐

Extremely intense None or extremely bland

4. What did you **LIKE** about the **FLAVOR** of this sample? ________

____

5. What did you **DISLIKE** about the **FLAVOR** of this sample? ________

____

6. Indicate by placing a mark in the box your **OVERALL LIKE/DISLIKE** for the **TENDERNESS** of this sample.

☐ ☐ ☐ ☐ ☐ ☐ ☐ ☐ ☐

Like extremely Neither Like nor Dislike Dislike extremely

7. Indicate by placing a mark in the box how you feel about the **LEVEL OF TENDERNESS** of this sample.

☐ ☐ ☐ ☐ ☐ ☐ ☐ ☐ ☐

Extremely tender Extremely tough

8. Indicate by placing a mark in the box your **OVERALL LIKE/DISLIKE** for the **JUICINESS** of this sample.

☐ ☐ ☐ ☐ ☐ ☐ ☐ ☐ ☐

Like extremely Neither Like nor Dislike Dislike extremely

9. Indicate by placing a mark in the box how you feel about the **LEVEL OF JUICINESS** of this sample.

☐ ☐ ☐ ☐ ☐ ☐ ☐ ☐ ☐

Extremely juicy Extremely dry

Juiciness is related to beef quality. Consumers rate dry meat lower for overall like/dislike and juicy meat higher for overall like/dislike. In a large in-home consumer study conducted in four cities (Philadelphia, Chicago, Houston, and San Francisco) where 300 consumer households were sampled, overall like/dislike of beef steaks was highly correlated to juiciness of the steaks (r = 0.77; Neely et al., 1998). Therefore, juiciness is highly related to overall acceptability of beef steaks and is an important palatability attribute.

Tenderness

Tenderness is defined as the ease with which a cube of meat is segmented by the molars during the chewing process. Tenderness is broken down into three sensory attributes by AMSA (1995): muscle-fiber tenderness, connective tissue amount, and overall tenderness. Muscle-fiber tenderness is the ease with which the muscle fibers segment and are broken down during chewing. A good example of muscle-fiber tenderness is the ease with which a fully cooked beef roast will segment and break down during chewing. This would most likely be rated an 8 on the ballot defined in Exhibit 7.1. Another example of an 8 or beef that would be extremely tender would be a beef tenderloin steak cooked to medium-rare degree of doneness. This steak would be from a Choice beef steer that was less than 20 months of age and had been fed a high concentrate diet for a minimum of 70 days. The steak would have been cooked using high heat on a grill. A 1 for muscle-fiber tenderness, or extremely tough in Exhibit 7.1, would be a steak that was derived from a muscle that had been placed in the freezer immediately postharvest (within 45 minutes post-exsanguination), thawed and cooked on a grill to medium degree of doneness. The process of freezing pre-rigor meat and thawing is called thaw rigor. The steak will shrink during thawing, which is the result of contraction within the muscle fibers. This process severely toughens a steak.

Connective tissue amount is defined as the amount of connective tissue remaining in the beef that is detectable during chewing and after the sample is fully chewed or the muscle fibers are fully broken down. This connective tissue amount does not include visible gristle or connective tissue that would normally be removed and not consumed. During chewing, connective tissue will not break down appreciably.

The most effective method of detecting connective tissue is after the sample is fully chewed, the sample is placed on the center of the tongue in a somewhat consolidated mass, and the tongue is moved through the sample. Connective tissue will be detected as strings or spider-web-type components or as fragments that did not break down during chewing. The beef tenderloin steak described earlier would most likely be an 8 for connective tissue amount as defined in Exhibit 7.1 and a grilled beef shank steak would be a 1 on this same scale.

Overall beef tenderness is usually the average between muscle-fiber tenderness and connective tissue amount. The exception to this is when connective tissue is an 8 or a 7 (Exhibit 7.1); then the overall tenderness is the same as the muscle-fiber tenderness. If there is no connective tissue or connective tissue is not contributing to the overall tenderness of the beef sample, then muscle-fiber tenderness would be the driving force for overall beef tenderness. If when averaging connective tissue and muscle-fiber tenderness, the resultant value is between a whole number (5.5), then the sensory panelist is to make a judgment as to the value that represents the overall tenderness of the sample (rate the sample a 5 or a 6).

Beef tenderness is an important component of overall consumer acceptability for beef. Lorenzen et al. (2003) reported that overall beef tenderness as rated by a trained sensory panel was moderately correlated ($r = 0.24$ in top loin steaks; $r = 0.15$ in top sirloin steaks; $r = 0.14$ in top round steaks) with overall consumer like/dislike. When consumers rated steaks as tough, they also did not like the steaks ($r = 0.85$; Neely et al., 1998). On the other hand, steaks that consumers rated as tender were more acceptable overall to consumers. Therefore, tenderness is considered an important component of beef palatability.

Flavor

Beef flavor is a very complex sensory attribute that is composed of multiple components. Flavor has been defined as the sum of the perceptions resulting from stimulation of the sense ends that are grouped together at the entrance of the alimentary and respiratory tracts (Amerine, Pangborn, and Roessler, 1965). Meilgaard, Civille, and Carr (1999) used the definition of Caul (1957) where flavor was defined as the impressions perceived via the chemical senses from a product in the mouth. Using the definition of Meilgaard, Civille, and Carr (1999), beef flavor includes flavor aromatics, basic tastes, and chemical feeling factors. Flavor aromatics are volatile compounds sensed through the olfactory bulb and are compounds released during chewing. Basic tastes are salty, sweet, sour, and bitter. Basic tastes are sensed on the tongue and are the result of soluble compounds being detected when the product is placed in the mouth and chewed. Chemical feeling factors are detected by the soft membranes of the mouth and nasal cavities and include astringency, spice heat, cooling, bite, metallic, and umami.

Beef flavor has been measured different ways. In AMSA (1995), overall flavor is rated where 1 = extremely bland and 8 = extremely intense (Exhibit 7.1). Panelists then rate any off-flavor by either identifying the flavor attribute using predetermined attributes or writing a descriptor in the off-flavor category if a descriptor is not defined. Panelists may or may not quantify the intensity of the off-flavor attribute. While this method provides some information on flavor attributes, it is difficult to quantify and differentiate intensity differences of attributes. While the AMSA (1995) method is widely used, if flavor differences are important to assess, it is not the recommended sensory method for assessing flavor in beef. The Spectrum and QDA sensory methods provide a much more thorough and quantitative method of assessing flavor, and this author prefers use of the Spectrum Method. Using the Spectrum Descriptive Attribute sensory method, Table 7.1 provides a baseline list of flavor attributes for beef. Exhibit 7.2 provides a sensory ballot that incorporates the beef flavor lexicon for determining beef flavor differences in a study. Note that the ballot in Exhibit 7.2 has additional flavor attributes listed that are not included in Table 7.1. The sensory panelists used the baseline lexicon and added flavor and after-flavor attributes to the ballot that were specific to the project. These attributes were derived from personal experience and using the flavor attributes defined in Civille and Lyons (1996). It is obvious when comparing the flavor aspects of the ballots in Exhibit 7.1 versus Exhibit 7.2 that beef flavor attributes are more thoroughly identified and quantified using the Spectrum Method. The QDA sensory method also assesses flavor. As seen in Figure 7.1, overall flavor aroma was assessed along with specific attributes of flavor. In the QDA method, panelists use their own descriptors of flavor and the computer will group similar flavor attributes into an attribute reported as a spoke on the spiderweb. In this sensory method, flavor is fully described and quantified.

Flavor has been shown to be of similar importance to consumers as beef tenderness. Neely et al. (1998) found very high correlation between consumer ratings for overall flavor desirability and beef flavor intensity with overall consumer acceptability (r = 0.86 and r = 0.79, respectively). These data indicate that flavor is an important component of consumer overall acceptability for beef.

FACTORS IMPACTING BEEF SENSORY ATTRIBUTES AND QUALITY

Preharvest

Cattle genetics, management, diet, handling, environment, and the stress susceptibility or adaptability of an animal impact meat sensory attributes. These factors may individually influence meat palatability or they may interact or influence one another.

Extensive research to understand genetic influences on beef palatability has been conducted. It is generally accepted that beef palatability traits are moderately heritable. Genetic differences between and within breeds have centered on differences in beef tenderness and marbling. Research at the Roman L. Hruska U.S. Meat Animal Research Center in Clay Center, Nebraska, has been conducted to evaluate differences in beef palatability between multiple breeds (Table 7.2). In these studies steers are harvested at a constant fat thickness in order to produce steers from each breed-type that represents industry practices. Differences in flavor and juiciness were minimal between breeds, but tenderness differences, either using Warner-Bratzler shear force or sensory tenderness, were found. Differences between breeds were minimal except when *Bos indicus* breed-types were compared to *Bos Taurus* breed-types. *Bos indicus* breed- types had higher Warner- Bratzler shear force values, or were tougher, than *Bos Taurus* breed-types. Research has further documented that as the percentage of *Bos indicus* breeding increases, beef tenderness tends to decrease and the variability in tenderness increases (Table 7.3 as adapted from Shackelford [1992]). The difference in beef tenderness between *Bos indicus* versus *Bos taurus* has

TABLE 7.2. Summary of Warner-Bratzler shear force (kg) (WBS) and beef palatability traits (flavor, juiciness, and tenderness) between beef cattle breed-types evaluated in the Germ Plasm Evaluation program at the USDA, ARS Roman L. Hruska U.S. Meat Animal Research Center in Clay Center, Nebraska, for Cycle I, II, and III.

Breed group	WBS	Flavor	Juiciness	Tenderness
Jersey	3.1	7.5	7.5	7.4
Hereford × Angus	3.3	7.3	7.3	7.3
South Devon	3.1	7.3	7.4	7.4
Sahiwal	4.1	7.1	7.0	5.8
Brahman	3.8	7.2	6.9	6.5
Brown Swiss	3.5	7.4	7.2	7.2
Gelbvieh	3.5	7.4	7.2	6.9
Simmental	3.5	7.3	7.3	6.8
Maine-Anjou	3.4	7.3	7.2	7.1
Limousin	3.5	7.4	7.3	6.9
Charolais	3.3	7.4	7.3	7.3
Chianina	3.6	7.3	7.2	6.9

Source: Adapted from GPEP (1974, 1975, 1978) and Dikeman (2003).

been contributed to *Bos indicus* cattle having higher levels of calpastatin, the protein that regulates the ability of calpains to degrade, or age, beef postharvest. If cattle have higher levels of calpastatin, their meat will be tougher early postharvest and the aging process will be slower. Figure 7.2 shows a typical decline or improvement of meat tenderness with aging. The steaks from the Angus or Hereford × *Bos indicus* steers were tougher on day one postharvest when compared to the tenderness of steaks from the Angus steers. With increased length of postharvest aging, Warner-Bratzler shear force values declined for both genetic-types. After 14 days of postharvest aging, steaks from the Angus or Hereford × *Bos indicus* steers had similar tenderness to steaks from the Angus steers. The steaks from the Angus steers improved in tenderness from one to seven days, but substantial improvements in tenderness were not seen after seven days. This indicates that while the Angus or Hereford × *Bos indicus* steers were initially tougher, they required up to 14 days of postharvest aging in order to maximize the effect of aging on beef tenderness whereas the meat from the Angus steers only required seven days of

TABLE 7.3. Warner-Bratzler shear force (kg) means from the *Longissimus* muscle of cattle differing in *Bos indicus* versus *Bos taurus* inheritance.

		Percentage *Bos indicus* breeding						
Reference	**Breed**	**0**	**25**	**38**	**50**	**62**	**75**	**100**
Damon et al., 1960	B	6.22	6.72	6.68	7.14	–	7.79	9.27
Carpenter et al., 1961	B	–	3.93	–	5.02	–	4.65	5.29
Ramsey et al., 1963	B	2.31	–	3.03	2.46	–	–	3.23
Luckett et al., 1975	B	3.94	4.37	–	–	–	–	6.30
Koch et al., 1982	B	3.44	–	–	3.92	–	–	–
Koch et al., 1982	S	3.44	–	–	4.27	–	–	–
McKeith et al., 1985	B	4.79	–	–	5.77	–	–	7.18
Bidner et al., 1986	B	3.90	4.30	–	–	–	–	–
Riley et al., 1986	B	4.30	–	–	6.00	–	–	–
Crouse et al., 1987	B	4.00	–	–	7.50	–	–	–
Crouse et al., 1987	S	4.00	–	–	8.00	–	–	–
Crouse et al., 1989	B	4.40	5.16	–	5.80	–	6.68	–
Crouse et al., 1989	S	4.40	5.64	–	6.64	–	8.41	–
Cundiff et al., 1990	N	5.50	–	–	7.00	–	–	–
Wheeler et al., 1990	B	4.75	–	–	4.75	–	–	6.40
Whipple et al., 1990	S	4.70	–	6.40	–	7.70	–	–
Shackelford et al., 1991	B	4.50	–	–	–	5.40	–	–

Source: Adapted from Shackelford (1992).

Note: B = Brahman; S = Sahiwal; and N = Nelore.

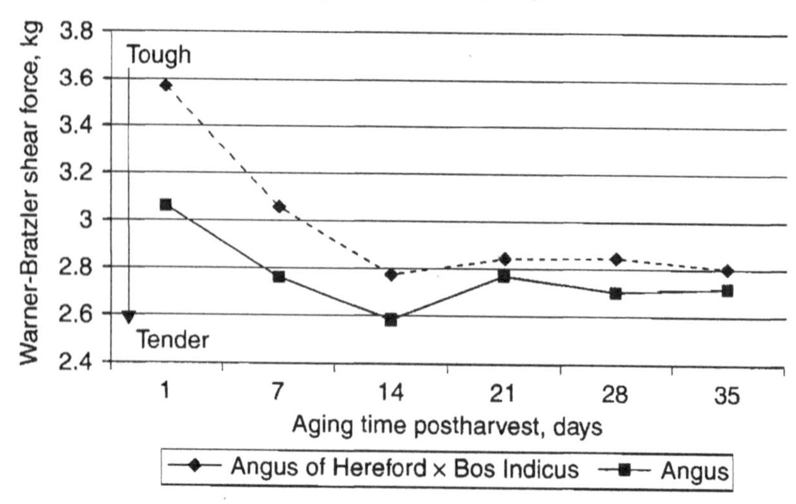

FIGURE 7.2. Effect of aging time postharvest on Warner-Bratzler shear force values in Angus and Hereford or Angus × *Bos indicus* steers.

aging to maximize beef tenderness. Interestingly, the 24-hr calpastatin level for meat from the Angus or Hereford × *Bos indicus* steers was 2.56 activity/gm and from the Angus steers it was 2.43 activity/gm. While not appreciably different, there was a tendency for meat from Angus or Hereford × *Bos indicus* steers to have higher 24-hr calpastatin levels. Differences in calpastatin levels have been mainly used to explain tenderness differences between *Bos indicus* and *Bos Taurus* cattle.

Variation in tenderness exists within a breed indicating that selection for tenderness would improve overall tenderness. Recent research has concentrated on the development of single gene markers for selection of cattle to improve tenderness. The GeneSTAR test for tenderness is being marketed by Genetic Solutions. This test examines variants of the calpastatin gene. The test rates cattle based on how many stars they have; 0, 1, or 2. They predict that cattle with 0 stars will have Warner-Bratzler shear force values 0.37 kg higher than cattle with 2 stars and that unacceptable consumer tenderness ratings will be reduced by 50 percent in 2-star cattle. Other gene markers tests are being developed. Page et al. (2002) discovered a single nucleotide polymorphism of the *m*-calpain gene. While research continues to more fully elucidate the genetic regulation of beef tenderness, most likely beef tenderness is not regulated by a single gene, but multiple genes. Selection for one gene may not be sufficient to make significant improvements in beef tenderness, but improved knowledge and understanding of the genetic regulation of beef palatability, specifically beef tenderness, will help beef producers understand how to select and control genetic effects to ultimately improve beef palatability.

Dietary influences on meat quality have been extensively studied and it has been shown that intensive feeding of high-concentrate diets prior to slaughter positively affects beef sensory properties (Meyer et al., 1960; Hawrysh, Berg, and Howes, 1975; Kropf, Allen, and Thouvenelle, 1975; Bowling et al., 1977; Schroeder et al., 1980; Tatum, 1981). Beef palatability improvement with feeding high-concentrate diets has been associated with multiple factors. First, overall carcass fatness, muscle mass, and carcass weight are increased. Also, increased time on high-concentrate diets increases marbling scores and sensory panel palatability ratings (Greene, Bacus, and Riemann, 1989; Williams, Keele, and Waldo, 1992; May et al., 1992). When cattle are fed high-concentrate diets, they grow more rapidly and they reach slaughter weight in a shorter period of time so they are younger. Therefore, the negative effects of increased age on meat palatability are diminished. As fed cattle are heavier with carcasses containing higher amounts of subcutaneous fat and greater muscle mass, they are not as susceptible to cold shortening. Also, cattle fed high-concentrate diets grow at rapid rates and increased growth rates have been shown to improve beef tenderness through increasing the solubility of the meat collagen (Aberle et al., 1981; Wu et al., 1981). Feeding cattle high-concentrate diets prior to slaughter also has been associated with removing variation associated with feeding effects prior to the cattle being fed high-concentrate diets. The lower the energy density of the diet, the more restricted the cattle growth. Cattle entering the feedlot that have been fed varying energy-based diets from high to low energy would expectantly differ in live weight and composition and, therefore, beef palatability would expectantly vary at this time. By feeding high-concentrate diets, this variation is reduced (Harrison et al., 1978; Skelley et al., 1978; Schroeder et al., 1980; Miller et al., 1987).

Feeding cattle high-concentrate diets also is related to improving beef flavor and juiciness. The diet can affect overall fatness level and so the affect of flavor may be due to changes in fat content. Meat with lower fat content is often described as being more beefy or brothy, higher in serumy, bloody, livery, and grainy/cowy flavors, and is more metallic than beef with higher amounts of fat. Fat most likely masks the other flavor attributes by slightly coating the mouth and decreasing ability to detect other flavor attributes. β carotene derived from forages may be deposited in fat and results in more yellow fat that is visually undesirable. Other off-flavors deposited in the adipose tissue also can be derived from dietary forage sources. Melton (1983) summarized that the corn in beef high-concentrate finishing diets can be partially or totally replaced by corn silage, a combination of corn silage and alfalfa, alfalfa hay, or a combination of alfalfa hay and timothy, and beef flavor would not be affected. Changes in the grain sources within a high-concentrate diet most likely will not affect beef flavor. Miller et al. (1997) fed beef steers either a corn-, or a corn/barley-, or a barley-based high-concentrate finishing diet 102 to 103 days prior to slaughter, to a final live weight of approximately 495 kg. Cooked top loin steaks did not differ in cooked beef flavor intensity or in any beef flavor attribute due to grain source of the diet.

The meat from beef fed corn-based diets can differ in flavor from pasture-fed beef. Most of the flavor difference is due to fatness differences in the beef (Melton, 1983). When pasture- and grain-fed cattle are slaughtered at similar fatness, Melton (1983) found that beef from pasture-fed cattle was still less desirable. These differences were most likely due to deposition of feed-derived compounds deposited in the fat. Supplementing cattle with grain during pasture feeding will dilute out these effects or feeding cattle for 90 to 100 days on grain-based diets prior to slaughter will reduce the negative flavor effects that may be induced by forages.

The environment that cattle are exposed to or are managed in may or may not impact meat palatability. As the environment impacts live animal growth, beef palatability may be affected. In addition, the nutrition that is available and the breed-type of the cattle may influence this relationship. In general, environments that do not limit live animal growth and provide sufficient nutrition for growth do not influence meat palatability. However, if the environment limits live animal growth through extremes in temperature and/or nutrient availability, beef palatability will most likely be affected through a reduction in live animal weight at harvest. Lighter cattle at harvest will have lighter carcass weights, carcasses with less subcutaneous fat, and lower amounts of marbling. Heat stress has been shown to negatively impact live animal performance. Heat-stressed steers had lower average daily gains, lower dry matter intake, and decreased feed efficiency when compared to steers that received shading to mitigate heat stress (Mitlohner et al., 2001; Mitlohner, Gaylean, and McGlone, 2002). In addition, some breed-types are more heat-tolerant. It is well documented that *Bos indicus*-influenced cattle have greater heat tolerance and are not as cold-tolerant when compared to other cattle types. In conclusion, fitting the breed-type with the environment and providing sufficient nutrition within an environment are the main factors to assuring that environment does not negatively impact beef palatability.

The effect of live animal temperament on beef palatability has been examined. The effect of animal temperament and the animal's subsequent response and ability to adapt to stressors or handling has been related to live animal performance (Mitlohner, Gaylean, and McGlone, 2002). It has been hypothesized that repeated activation of the hypothalamic–pituitary–adrenal axis during growth would have negative impacts on carcass characteristics and meat tenderness. Cattle with more excitable temperaments have been reported to produce less tender beef than cattle with calmer temperaments (Voisinet et al., 1997). Brown et al. (2004) and Vann, Paschal, and Randel (2004) showed that flighty steers in response to handling (the stressor) had lower average daily gains, higher feed-to-gain ratios, lower USDA Quality and Yield grades, and higher Warner-Bratzler shear force values. In Santa Gertrudis, with cattle from the King Ranch, Carstens et al. (2005) found that flightier cattle at the end of the finishing period just prior to harvest were tougher than calmer cattle ($r = 0.37$ and 0.27 for exit velocity with 1-day and 14-days Warner-Bratzler shear force). Falkenberg et al. (2005) reported that flightier steers at weaning had higher Warner-Bratzler shear force values than calmer steers (2.83 lbs. versus 2.46 kg, respectively). Some cattle adapt or cope with stress more effectively than others. Identification or measurement of cattle's response to stress may provide a tool for identification of cattle that are not negatively impacted by stress. Collectively, results from these studies demonstrate that temperament classification of calves may be a useful management tool to sort feeder calves into outcome groups that differ in growth performance and beef tenderness traits.

The age of animals at harvest has been shown to impact meat palatability. Meat from very young animals, veal for example, has been shown to be tender and to not have as much flavor as meat from older animals, such as 18-month-old steers produced from a high-concentrate finishing diet. As animals age, flavor changes and meat becomes tougher due to the increased amount and decreased heat-solubility of the connective issue (called perimysium and endomysium) within the muscle. Perimysium is the connective tissue that surrounds muscle-fiber bundles within the meat and endomysium is the connective tissue that surrounds individual muscle fibers. Meat from older cattle or C, D, or E maturity carcasses, is more intense in flavor due to the meat having inherently higher amounts of myoglobin that results in higher serumy/bloody, livery, cowy, and metallic flavor attributes. In addition, this meat also is tougher due to a higher amount of connective tissue, and the connective tissue that is present does not break down with heat, or it is less heat-soluble. During heating or cooking, heat-soluble bonds break down and the resultant meat is more tender. When a higher proportion of the collagen bonds are heat-stable, during cooking these bonds are not broken and the resultant meat is tougher. This change in the heat solubility of the connective tissue is one of the main reasons that maturity is included in the USDA Beef Quality Grading system (Chapter 6).

In conclusion, it is obvious that preharvest factors influence beef palatability. As discussed, genetics, environment, diet, stress susceptibility or temperament, and animal age influence the eating quality of beef. While additional research is needed to fully understand how each of these factors influences beef palatability, the interaction of these factors on beef-eating quality also needs to be elucidated.

Postharvest

At the point of harvesting, during harvesting, chilling, and storage, beef sensory attributes can be influenced positively or negatively. The effects of immobilization and stunning, electrical stimulation, chilling, pH decline, postharvest aging, blade tenderization, and product enhancement are the main postharvest factors that have been shown to affect beef palatability.

The type and application of the immobilization method used during harvesting can impact meat quality either through inducing short-term preslaughter stress or it can affect blood removal upon exsanguinations. During immobilization, the animal should be rendered unconscious without stopping the action of the heart. The purpose of immobilization is to reduce animal suffering. For cattle, concussive methods are used for immobilization. A concussive force is applied to the head of the animal to render it unconscious and to eliminate struggling. If animals are not rendered completely unconscious, animals may experience short-term stress that results in glycogen-depletion of muscle and increased muscle metabolism. This can result in a more rapid than normal pH decline postharvest and the meat will be pale, soft, and exudative (PSE). Short-term stress also can result in a meat quality defect called ecchymosis or blood splash. Ecchymosis is the result of short-term stress while the heart is functioning. When the animal becomes stressed, blood pressure increases, which results in hemorrhaging within the capillaries of the muscle and fat tissues. The meat and fat will have blood spots that are visibly undesirable.

During exsanguinations, proper and adequate blood removal can be affected by proper application of the immobilization technique. If an animal experiences short-term stress as a result of improper immobilization, blood removal from the extremities may be limited. As a result the meat will contain higher amounts of blood that will result in higher levels of serumy/bloody, cowy, livery, and metallic beef flavor attributes. Inadequate severing of the artery or vein used for exsanguination, extended time between stunning and exsanguination, and improper suspension of the carcass during exsanguination can restrict the volume of blood removed. When hemoglobin content is higher in muscle tissue due to improper bleeding, ecchymosis may occur, the meat will be darker red, and have higher levels of serumy/bloody, cowy, livery, and metallic beef flavor attributes. Proper application of immobilization and exsanguination techniques is important in maximizing beef palatability.

Electrical stimulation is a method of applying an electrical pulse to a beef carcass during the harvesting process after immobilization and exsanguination. This electrical pulse is cycled off and on. During the application of the electrical pulse, muscles contract and when the pulse is turned off, muscles relax. During contraction, muscles use up available energy. The cycling of the electrical pulse, or the contraction and relaxation cycle of the muscles, uses up the available energy stores within muscles early postharvest. This forces muscles to go into rigor mortis more rapidly than nonstimulated muscles. As a result, when beef carcasses are placed in a cooler, the muscles are not susceptible to a condition called "cold-shortening." Cold-shortening occurs when muscles are chilled, but still have available energy and they have not progressed far into rigor mortis. As muscles are chilled, the mechanisms that control the amount of free calcium are lost, free calcium levels increase, and muscles contract more than normal or shorten. Since the cold temperature is what induced the inability of the muscle to control free calcium, this process is called "cold-shortening" or "cold-induced shortening." When muscles contract, proteins that provide structural support to the outer portion (called Z lines) of the muscle's smallest contractile apparatus (called a sarcomere) are concentrated. These main proteins or Z lines are very strong and when they are concentrated, beef is tougher. Therefore, electrical stimulation uses up the energy in the muscle before it is chilled and does not allow the process of "cold-shortening" to occur. Electrical stimulation can be applied in varying voltages and is generally classified as low-voltage, medium-voltage, and high-voltage. Low-voltage electrical stimulation (80V or less) is applied very early in the harvesting process. Many times, low-voltage electrical stimulation is applied immediately post-exsanguination to assist with proper blood removal and to assist in reducing cold-shortening. Medium- and high-voltage electrical stimulation is applied to the beef carcass during the harvesting process and prior to chilling. Carcasses that have less than 10 mm of exterior or subcutaneous fat and light weight carcasses (less than 250 kg) are more susceptible to "cold-shortening" due to the reduction in the amount of insulation (exterior fat) and muscle mass. These beef carcasses chill rapidly and electrical stimulation significantly improves carcass tenderness. An added benefit is that electrical stimulation does not negatively affect beef carcasses that have sufficient muscle and carcass mass and exterior fat. Electrical stimulation can be used on all carcasses without deleterious effects and it will improve tenderness for carcasses that are susceptible to "cold-shortening."

pH decline of beef muscles is an indication of how a muscle is progressing through rigor mortis. How rapidly or slowly a muscle proceeds through rigor mortis affects the final pH of the muscle, can affect the contractile state of the muscle (if it is cold-shortened or not), and may influence meat tenderness. Researchers have examined the effect of pH three hours postharvest on beef tenderness. Jones and Tatum (1994) found that beef muscles that had pH values less than 6.2 at three hours postharvest were more tender (based on Warner-Bratzler shear force values) than muscles with pH values greater than 6.46 at three hours in electrically stimulated beef carcasses. Marsh et al. (1987) found that beef muscles with intermediated pH at three hours (pH of about 6.1) had the highest tenderness values. These muscles were defined as having an intermediate glycolytic rate (defined as the rate in which glycolysis proceeded). Marsh et al. (1987) included nonelectrically stimulated and electrically stimulated beef carcasses and four different chilling routines to induce differences in pH decline. They concluded that muscles with rapid and slow glycolytic rates were tougher and that this toughening effect was sustained for up to 14 days of postharvest aging. Smulders et al. (1990) also examined the effect of pH decline postharvest on beef tenderness in electrically stimulated beef carcasses. They found that steaks from beef carcasses with three-hour pH values below 6.3 were more tender and less variable in tenderness than steaks from beef carcasses with three-hour pH above 6.3. Pike et al. (1993) reported that steaks from carcasses with three-hour pH values of 6.0 had optimal tenderness. Shackelford, Koohmaraie, and Savell (1994), using a diverse and large set of cattle ($n = 444$) that included electrically stimulated and non-electrically stimulated carcasses, did not find a relationship between three-hour pH and beef tenderness. While the research is conflicting, pH decline postharvest has been shown to influence beef tenderness in some studies. When applying three-hour pH measurements within a commercial application, three-hour pH may not be an effective predictor of pH due to increased variation in other factors that may affect pH decline. Postharvest handling, minimal stress during exsanguination and proper length of time between immobilization and exsanguination, length of time for harvesting, electrical stimulation application and type, chill temperature, and muscle metabolism also may affect pH decline. If animals are excited prior to harvesting or during immobilization or exsanguination, their metabolism is more rapid and this results in a more rapid pH decline. If carcasses are exposed to elevated temperatures for longer than normal times, pH will drop more rapidly than normal. Elevated temperatures are related to delayed time on the harvesting floor due to either mechanical breakdowns that stop the harvesting process or carcasses that are segmented off the processing line for inspection or safety concerns.

Chilling, as previously discussed, can negatively influence beef tenderness through inducing "cold-shortening." If electrical stimulation is utilized during harvesting, this effect is mitigated. Chilling also has an effect on beef palatability in that even, consistent, chilling of carcasses is necessary so that the effect of temperature on pH decline is consistent across muscles within a carcass. Chilling rate also has been implicated as influencing beef tenderness through the effect that temperature has on the rate of glycolysis or glycogen depletion. Marsh et al. (1987) showed that electrically stimulated carcasses that were chilled slowly produced tough meat. They hypothesized that the rapid rate of glycolysis induced by electrical stimulation in combination with the muscle being maintained at an elevated temperature resulted in a high rate of pH decline and induced muscle toughening. The exact mechanism responsible for the toughening effect was not known. It is obvious that consistent and rapid chilling of muscles is important for beef safety and the chilling rate can influence beef tenderness.

The length of time that the meat is stored in refrigerated temperatures postharvest is defined as postharvest aging. Aging has a significant effect on beef tenderness and can influence beef juiciness and beef flavor. As meat is aged, enzymes that are inherent in the muscle break down the protein structure within the meat. This results in improved beef tenderness. The enzymes that work to break down the structural proteins are called "calpains." These enzymes work in living muscle to help repair muscle proteins and they are inherently part of the muscle cell. There is also a regulator of calpains called calpastatin. Calpastatin is very important in living muscle in that calpains would always be active and muscle proteins would be continuously degraded without the controlling mechanism of calpastatin. The level of calpastatin has been related to how rapidly meat ages. If a muscle has a higher level of calpastatin, the aging process is slower and if a muscle has a lower level of calpastatin, the meat ages or improves in tenderness more rapidly. Figure 7.2 shows a typical decline or improvement of meat tenderness with aging. Figure 7.2 shows that meat from different animals age at different rates and improvement in meat tenderness is obtained for most meat within 14 days postharvest.

During aging, beef flavor can be affected. With increased storage time, flavors associated with microbial growth can increase. In vacuum-packaged beef, soured flavors associated with growth of acid-producing bacteria can increase. In aerobically stored beef, putrid, rotten-egg, and sulphur-based flavors increase with increased storage due to the growth of gram-negative spoilage bacteria. Beef flavor also can be affected by the oxidation of lipids. These flavor changes have been described as rancid or old. Johnsen and Civille (1986) defined flavors associated with lipid oxidation as cardboardy, painty, fishy, and livery (Table 7.1). Increased levels of metallic and astringency mouthfeels also can be associated with increasing levels of lipid oxidation. Aging beef postharvest is a balancing act between aging long enough to improve beef tenderness, but not aging too long to result in increased levels of microbial growth and lipid oxidation. Vacuum-packaged beef with low microbial levels upon packaging should enable storage for 14 to 21 days to improve beef tenderness without resulting in microbial levels and lipid oxidation that would negatively affect beef flavor or safety.

Blade tenderization is a production practice that has been used for many years in the beef industry to improve tenderness of beef. Beef cuts are passed through a line of thin blades that puncture the meat or needle it. This process disrupts the muscle fibers and the perimysial and endomysial connective tissue. Many researchers have shown an improvement in beef tenderness from 0.2 to 1 kg in Warner-Bratzler shear force when beef cuts are blade tenderized (Savell, Smith, and Carpenter, 1977; Tatum, Smith, and Carpenter, 1978). Blade tenderization should occur at the end of the postharvest aging period as blade- tenderized beef will have excessive purge and be more susceptible to microbial spoilage with extended storage even in a vacuum-package.

Another technology that is currently being used in the beef industry to improve beef juiciness and/or tenderness, enhance flavor, improve color, stabilize color, increase shelf-life, improve safety, or increase water-holding capacity in the final product is the addition of an injection solution containing nonmeat ingredients, or meat enhancement. The use of injections or the addition of ingredients to meat is a historically old practice. Water is used as a carrier and ingredients such as salt, sodium phosphates, sodium or potassium lactate, and sodium diacetate are used for their functional properties. Salt is added to beef to increase shelf-life, enhance flavor, and increase the water-holding capacity of the meat proteins that results in decreased purge loss and improved cook yields. Salt or sodium chloride acts to improve water-holding capacity through inducing the swelling of meat proteins (Hanson and Huxley, 1955). Offer and Trinick (1983) found that meat proteins have the ability to swell to twice their size in salt concentrations. The swelling provides a higher number of protein side chains that can bind water and, therefore, water-holding capacity is increased. Sodium chloride addition also enhances meat flavor and has been associated with increasing basic salt taste and increasing lipid oxidation.

Sodium chloride is often added in combination with sodium phosphates. The challenge is to balance the amount of sodium chloride addition with the amount of sodium phosphate added to maximize water-holding capacity without getting too high a salt flavor or altering texture so that the meat is too soft or has a more processed-meat-like bite. Sodium phosphates were originally used in meat-injection solutions to enhance water-holding capacity. Their usage is regulated by USDA, FSIS, and fresh meat products cannot contain more than 0.5 percent in the final product (Federal Register, 1982). There are various forms of sodium phosphates that can be added to meat products; however, sodium tripolyphosphate is the predominant phosphate in phosphate blends used in meat systems. The major sodium phosphates purchased commercially contain blends of different phosphates and these blends are used to increase functionality. Sodium tripolyphosphate has an alkaline pH and even though meat is a very good buffer, the addition of sodium tripolyphosphate to meat systems increases meat pH. Increasing meat pH improves water-holding capacity by moving the meat pH further from the meat protein isoelectric point. As the pH moves further from the isoelectric point (about 5.2 to 5.3 in red meat), water-holding capacity increases due to an increase in the amount of negative charges on the meat proteins that can bind water (Trout and Schmidt, 1986). The net result is an increase in the amount of water that can be bound to the meat proteins.

With increased water-holding capacity, meat containing sodium phosphates has been shown to be juicier, have improved tenderness, and have some changes in flavor. The addition of sodium phosphates alone is commonly associated with increasing soapy and sour off-flavors. Vote et al. (2000) reported increased flavor attributes with sodium phosphate addition to beef loin steaks. The addition of other ingredients in combination with salt and sodium phosphates, like sodium lactate, have been shown to help mask soapy and soured flavors associated with sodium phosphate addition (Vote et al., 2000).

Sodium or potassium lactate, the salt form of lactic acid, has been used as a nonmeat ingredient in the meat industry since the late 1980s. The initial work (Maas, Glass, and Doyle, 1989) showed that sodium lactate restricted development of botulinum toxin in cook-in-bag turkey products. Sodium lactate was approved for use at up to 2 percent as a flavor enhancer by USDA, FSIS (Federal Register, 1990), but it can be used at up to 4.0 percent on an individual basis if the company requests special consideration for its use as an antimicrobial agent. The addition of sodium lactate has been shown to increase meat juiciness (Papadopoulos, Miller, Ringer, and Cross, 1991; Eckert et al., 1997; Maca, Miller, and Acuff, 1997). As sodium lactate increases cook yield when used in combination with low salt brines, improvements in juiciness in meat containing sodium lactate are not surprising.

Sodium lactate has been shown to enhance the cooked beefy/brothy flavor aromatic and to limit the subsequent decline in this aromatic during refrigerated storage in beef products (Papadopoulos, Miller, Ringer, and Cross, 1991; Papadopolous, Miller, Acuff, et al., 1991; Evans, 1992; Pagach, 1992; Eckert et al., 1997; Maca, Miller, and Acuff, 1997; Maca, Miller, Maca, and Acuff, 1997; Maca et al., 1999). As beefy flavors declined during refrigerated storage, increased intensity of aromatics associated with lipid oxidation have been reported. Therefore, sodium lactate addition has been shown to improve the beefy flavor and to reduce the development of off-flavors associated with lipid oxidation. However, sodium lactate addition also has been associated with higher salt tastes, increased throat-burning mouthfeel, and higher levels of chemical aromatic flavor. These negative flavor attributes have been characterized as components of the sodium lactate itself. The addition of sodium lactate with other ingredients appears to decrease these effects and these effects are only slightly detectable at levels up to 3 percent. While sodium or potassium lactate addition improves beef palatability, these ingredients are added to beef enhancement solutions to improve beef safety and shelf life.

Sodium diacetate recently has been approved for use in meat products as an antimicrobial agent. It can be added alone, but many times it is added in combination with other antimicrobial agents such as sodium lactate or potassium lactate. Sodium diacetate is a compound containing 60 percent sodium acetate and 40 percent acetic acid. It is traditionally added at low levels in the final product, usually not greater than 0.2 percent in meat products, and more commonly, it is added at levels of 0.1 to 0.15 percent in the final product. The addition of sodium diacetate was shown to not affect the juiciness of ground beef patties (Grones, 2000) or enhanced beef strip loin steaks (Anwar, 2000). As sodium diacetate lowers meat pH, it could possibly decrease juiciness as the meat pH is moving closer to the meat isoelectric point, and a decrease in the water-holding capacity would be expected. However, the decrease in pH with sodium diacetate addition is most likely not sufficient enough to appreciably affect meat juiciness (Grones, 2000). Similarly, sodium diacetate addition does not appear to significantly affect meat texture or flavor as it is used at such low levels (Weber, 1997; Grones, 2000).

Another use of injection technology is the injection or infusion of calcium chloride into pre-rigor or meat within a short time postharvest to improve meat tenderness. Calpains are stimulated to work by free calcium, so inducing increased levels of free calcium through the injection or infusion of calcium postharvest should improve tenderness. The original research infused lamb carcasses with up to 10 percent of their body weight with a 0.3 M solution of calcium chloride immediately post-exsanguination and found that the calpain system was activated (Koohmaraie et al., 1988). The current recommendation is to inject up to 5 percent of the meat weight of a 2 M calcium chloride solution within one or two days postharvest and assure that the meat is aged seven days (Wheeler, Koohmaraie, and Crouse, 1991,

1992; Wheeler et al., 1993; Wheeler, Koohmaraie, and Shackelford, 1994). Wheeler et al. (1997) also showed that calcium chloride injection could be applied for up to 14 days postharvest, and improvements in tenderness were reported.

PALATABILITY OF DIFFERENT BEEF MUSCLE

The aforementioned discussion concentrates on the pre- and postharvest factors that affect beef palatability and is based on differences between animals or carcasses. However, significant beef palatability differences exist across muscles within a carcass. Muscles within a carcass differ in tenderness and flavor. Differences in muscles are due to the biological function of the muscle within the animal system. It has been well documented that muscles that are actively involved in the locomotion of the live animal have higher amounts of connective tissue, are tougher due to the connective tissue content, and may have higher myoglobin content than muscles used for structural support. Muscles used for structural support are more tender due to lower connective tissue content. Lipid content of muscles differs based on function of the muscles. Some muscles have higher amounts of glycolytic muscle fibers or white (Type IIB) muscle fibers. These fibers are larger in size, have lower myoglobin content, are more dependent on glycolysis for their energy source, and have low lipid content. Muscles that are needed for short bursts of speed or strength have a higher amount of white muscle fibers. As a result, these muscles are lower in myoglobin and lipid and have been characterized as having less cooked beef fat, serumy/bloody, cowy, livery, and metallic flavor attributes. Muscles that are needed for long, sustained work have a higher proportion of red muscle fibers (Type I). These fibers are smaller in diameter, darker red, have higher myoglobin content, use oxidative metabolism for their energy source, have lower amount of glycogen, and are higher in lipid content. Muscles that have a higher proportion of red fiber types have higher amounts of marbling and are redder in color. These muscles have higher amounts of cooked beef fat, serumy/bloody, cowy, livery, and metallic flavor component. Many muscles are intermediate or they contain a proportion of white, red, and intermediate muscle-fiber type. Intermediate muscle-fiber types (Type IIA) have characteristics intermediate to red and white fibers and have combinations of the flavor attributes defined for red and white muscles. In addition, the fatty acid content of the lipid in the muscle affects beef flavor. Differences in fatty acid content of meat have been associated with species-specific flavors or, in other words, lamb tastes different from beef due to the difference in the fatty acid profile of the lipids from these two species.

The National Cattlemen's Beef Association, the University of Nebraska-Lincoln, and the University of Florida conducted a study of individual beef muscles called Muscle Profiling. In this study, individual muscle characteristics including myoglobin content, connective tissue content, fatty acid profile, trained meat palatability as defined in Exhibit 7.1, Warner-Bratzler shear force, and many other attributes were measured. They showed that individual beef muscles differ in flavor. Other researchers have shown that beef muscles differ in flavor. McKeith et al. (1985) measured flavor desirability in 13 major beef muscles and then ranked them on flavor desirability (Table 7.4). There was a one-point difference using an eight-point scale in

TABLE 7.4. Rank of thirteen muscles by flavor desirability ratings where 1 = extremely undesirable and 8 = extremely desirable.

Muscle	Sensory score
Psoas major	6.55[a]
Infraspinatus	6.48[a]
Longissimus dorsi (loin)	6.37[a]
Longissimus dorsi (rib)	6.35[ab]
Rectus femoris	6.19[abc]
Biceps femoris	6.00[bcd]
Gluteus medius	5.87[cde]
Triceps	5.84[cde]
Supraspinatus	5.72[de]
Semitendinosus	5.66[de]
Semimembranosus	5.64[de]
Adductor	5.58[e]
Pectoral	5.52[e]

Source: Created based on data from McKeith et al. (1985).

[a,b,c,d,e] Means without a common superscript letter differ.

flavor desirability from the top- to bottom-ranked muscles. It is obvious that muscles differ in flavor. Contributions to differences in flavor between muscles have not been documented. Based on data collected for the USDA Nutrient Data (USDA, 1990), beef cuts differ in fatty acid profile (Table 7.5) and this difference most likely contributes to differences in flavor between muscles. Muscles also differ in total level of fat (Table 7.5), lean, and myoglobin content. All of these factors most likely contribute to flavor differences between muscles.

SUMMARY AND CONCLUSIONS

Beef sensory and palatability are important aspects of consumer acceptability. How beef palatability is measured impacts the interpretation of the results and what is measured. Trained Meat Descriptive Attribute Sensory Evaluation is a very useful sensory tool to measure the juiciness, tenderness, and overall flavor of beef. Flavor and Texture Descriptive Attribute Evaluation using either the Spectrum Method or the QDA method allow for a more thorough evaluation of beef flavor. Consumer sensory evaluation is the most effective tool to understanding the acceptability and preferences of consumers. These sensory tools have been used to understand beef palatability and the factors that affect beef palatability and ultimately consumer acceptability. Through the use of sensory tools, it has been established that animal genetics, the nutritional regimen of animals prior to harvest, the environment that cattle are produced in, live animal temperament or adaptability, and animal age impact beef palatability. Improvement and reduction in variation in beef-eating quality can be obtained through management and control of these preharvest factors. In addition, how animals are immobilized and exsanguinated, the use of electrical stimulation, pH decline, chilling rate, aging time postharvest, blade tenderization, enhancement, and calcium chloride injection are postharvest factors that can impact beef palatability. While pre- and postharvest factors impact palatability, the beef muscle and its function within the live animal also impact the eating quality of the subsequent beef. To maximize beef palatability, all of these factors must be considered and either controlled or used to maximize the acceptability for consumers of beef.

TABLE 7.5. Lipid and fatty acid composition of raw separable beef lean of different beef retail cuts for Choice beef grades.

	Beef Cut											
Lipid, grams	Chuck, arm	Chuck, blade	Rib, whole	Rib, large end	Rib, small end	Round, bottom	Round, eye	Round, tip	Round, top	Tender-loin	Top loin	Top sirloin
Total lipid	5.10	8.50	9.44	10.20	8.30	6.30	4.90	4.40	3.90	8.60	6.40	5.00
Total saturated	1.84	3.19	3.78	4.16	3.23	2.15	1.69	1.50	1.34	3.22	2.43	1.76
10:0	0.00	0.00	0.01	0.01	0.00	0.00	0.00	0.00	0.00	0.01	0.00	0.00
12:0	0.00	0.01	0.01	0.01	0.00	0.00	0.00	0.00	0.00	0.01	0.00	0.00
14:0	0.12	0.24	0.28	0.31	0.24	0.15	0.12	0.10	0.09	0.22	0.19	0.12
16:0	1.12	1.90	2.26	2.48	1.93	1.41	1.08	0.92	0.82	1.84	1.51	1.06
18:0	0.60	1.05	1.23	1.35	1.03	0.58	0.49	0.48	0.42	1.13	0.73	0.58
Total monosaturated	1.92	3.78	3.88	4.09	3.56	2.85	2.03	1.79	1.52	3.24	2.73	1.99
16:1	0.19	0.31	0.35	0.40	0.29	0.33	0.20	0.15	0.13	0.23	0.23	0.15
18:1	1.74	3.46	3.52	3.69	3.26	2.52	1.84	1.64	1.40	3.01	2.50	1.85
20:1	0.00	0.01	0.01	0.01	0.01	0.01	0.00	0.00	0.00	0.00	0.01	0.00
Total polyunsaturated	0.26	0.32	0.34	0.39	0.27	0.25	0.21	0.21	0.17	0.32	0.23	0.24
18:2	0.21	0.26	0.28	0.31	0.24	0.19	0.16	0.17	0.14	0.26	0.18	0.18
18:3	0.02	0.02	0.02	0.03	0.01	0.01	0.01	0.01	0.01	0.03	0.01	0.01
20:4	0.03	0.04	0.03	0.04	0.02	0.04	0.03	0.03	0.03	0.04	0.03	0.04

Source: USDA (1990).

LITERATURE CITED

Aberle, E.D., E.S. Reeves, M.D. Judge, R.E. Hunsley, and T.W. Perry. 1981. Palatability and muscle characteristics of cattle with controlled weight gain. Time on a high energy diet. *J. Anim. Sci.* 52: 757-763.

Amerine, M.A., R.M. Pangborn, and E.B. Roessler. 1965. *Principles of Sensory Evaluation of Foods.* Academic Press, New York, p. 549.

AMSA. 1978. Guidelines of Cookery and Sensory Evaluation of Meat. *Am. Meat Sci. Assn.* and National Live Stock and Meat Board, Chicago, IL.

———. 1995. Research Guidelines for Cookery, Sensory Evaluation and Instrumental Measurements of Fresh Meat. *Am. Meat Sci. Assn.* and National Live Stock and Meat Board, Chicago, IL.

Anwar, N. 2000. The effect of potassium lactate and sodium diacetate on the microbial, sensory, color and chemical characteristics of vacuum-packaged beef top loin steaks. MS thesis, Texas A&M University, College Station, TX.

Bidner, T.D., A.R. Schupp, A.B. Mohamad, N.C. Rumore, R.E. Montgomery, C.P. Bagley, and K.W. McMillin. 1986. Acceptability of beef from Angus-Hereford or Angus-Hereford-Brahman steers finished on all-forage or a high-energy diet. *J. Anim. Sci.* 62: 381-387.

Bowling, R.A., G.C. Smith, Z.L. Carpenter, T.R. Dutson, and W.M. Oliver. 1977. Comparison of forage-finished and grain-finished beef carcasses. *J. Anim. Sci.* 45: 209-215.

Brown, E.G., G.E. Carstens, J.T. Fox, K.O. Curley, T.M. Bryan. L.J. Slay, T.H. Welsh Jr., R.D. Randel, J.W. Holloway, and D.H. Keisler. 2004. Physiological indicators of performance and feed efficiency traits in growing steers and bulls. *J. Anim. Sci.* 82 (2): 54 (Abstr.).

Carpenter, J.W., A.Z. Palmer, W.G. Kirk, F.M. Peacock, and M. Koger. 1961. Slaughter and carcass characteristics of Brahman and Brahman crossbred steers. *J. Anim. Sci.* 20: 336-340.

Carstens, G.E., R.K. Miller, R.D. Randel, T.H. Welsh Jr., J.W. Holloway, L.O. Tedeschi, J. Pollack, D.P. Kirschten, J.J. Baker, D.G. Fox, P.C. Genho, and S.A. Moore. 2005. Determination of genetic and phenotypic variances of meat quality traits and their interrelationships with economically important traits in *Bos indicus* type cattle. Final Report to the National Cattlemen's Beef Association, June.

Caul, J.F. 1957. The profile method of flavor analysis. *Adv. Food Res.* 7: 1.

Civille, G.V. and B.G. Lyons. 1996. *Aroma and Flavor Lexicon for Sensory Evaluation: Definitions, References, and Examples.* ASTM Data Series Publication DS66. American Society for Testing Materials, W. Conshohocken, PA.

Crouse, J.D., L.V. Cundiff, R.M. Koch, M. Koohmaraie, and S.C. Seideman. 1989. Comparisons of *Bos indicus* and *Bos taurus* inheritance for carcass beef characteristics and meat palatability. *J. Anim. Sci.* 67: 2661-2668.

Crouse, J.D., S.C. Seideman, and L.V. Cundiff. 1987. The effect of carcass electrical stimulation on meat obtained from *Bos indicus* and *Bos taurus* cattle. *J. Food Qual.* 10: 407.

Cundiff, L.V., R.M. Koch, K.E. Gregory, J.D. Crouse, and M.E. Dikeman. 1990. Germ plasm evaluation program. Progress Report No. 12, pp. 1-6. Agricultural Research Center, USDA, U.S. Meat Animal Research Center, Clay Center, NE.

Damon, R.A., Jr., R.M. Crown, C.B. Singletary, and S.F. McCraine. 1960. Carcass characteristics of purebred and crossbred beef steers in the Gulf coast region. *J. Anim. Sci.* 19: 820-844.

Dikeman, M.E. 2003. Metabolic modifiers and genetics: Effects on carcass traits and meat quality. *49th Int'l. Cong. of Meat Sci. and Technol.* 49: 1-38.

Eckert, L.A., J.V. Maca, R.K. Miller, and G.R. Acuff. 1997. Sensory, microbial and chemical characteristics of fresh aerobically stored ground beef containing sodium lactate and sodium propionate. *J. Food Sci.* 62: 429.

Evans, L.L. 1992. L-Sodium lactate in cooked beef top rounds: Differing levels of incorporation and cookery. MS thesis, Texas A&M University, College Station, TX.

Falkenberg, S.M., R.K. Miller, J.W. Holloway, F.M. Rouquette Jr., G.E. Carstens, and R.D. Randel. 2005. Exit velocity effects on growth, carcass characteristics, and tenderness in half-blood Bonsmara steers. *Proceedings of the 51st Int'l. Cong. of Meat Sci. and Tech.* p. 29.

Federal Register. 1982. Rules and Regulations Vol. 47, No. 49, 10763.

———. 1990. Sodium lactate and potassium lactate as flavor enhancers and flavoring agents in various meat and poultry products. Federal Register 55: 7339.

GPEP. 1974. Progress Report: Germ Plasm Evaluation Program Report No. 1. U.S. Meat Animal Research Center, Clay Center, NE.

———. 1975. Progress Report: Germ Plasm Evaluation Program Report No. 2. U.S. Meat Animal Research Center, Clay Center, NE.

———. 1978. Progress Report: Germ Plasm Evaluation Program Report No. 6. U.S. Meat Animal Research Center, Clay Center, NE.

Greene, B.B., W.R. Bacus, and M.J. Riemann. 1989. Changes in lipid content of ground beef from yearling steers serially slaughtered after varying lengths of grain finishing. *J. Anim. Sci.* 67: 711-715.

Grones, K.L. 2000. Ground beef shelf-life assessment as influenced by sodium lactate, sodium propionate, sodium diacetate and soy protein concentrate addition to ground beef patties. MS thesis, Texas A&M University, College Station, TX.

Hanson, J. and H.E. Huxley. 1955. The structural basis of contraction in striated muscle. *Symp. Soc. Exp. Biol.* 9: 228.

Harrison, A.R., M.E. Smith, D.M. Allen, M.C. Hunt, C.L. Kastner, and D.H. Kropf. 1978. Nutritional regime effects on quality and yield characteristics of beef. *J. Anim. Sci.* 47: 383-388.

Hawrysh, Z.J., R.T. Berg, and A.D. Howes. 1975. Eating quality of beef from steers fed full or restricted levels of moisture treated barley. *Can. J. Anim. Sci.* 55: 179-185.

Johnsen, P.B. and G.V. Civille. 1986. A standardized lexicon of meat WOF descriptors. *J. Sensory Studies* 1: 99.

Jones, B.K. and J.D. Tatum. 1994. Predictors of beef tenderness among carcasses produced under commercial conditions. *J. Anim. Sci.* 72: 1492-1501.

Koch, R.M., M.E. Dikeman, and J.D. Crouse. 1982. Characterizations of biological types of cattle (Cycle III). III. Carcass composition, quality and palatability. *J. Anim. Sci.* 54: 35-45.

Koohmaraie, M., A.S. Babiker, A.L. Schroeder, R.A. Merkel, and T.R. Dutson. 1988. Acceleration of postmortem tenderization in ovine carcasses through activation of Ca^{+}2-dependent proteases. *J. Food Sci.* 53: 1638-1641.

Kropf, D.H., D.M. Allen, and G.J. Thouvenelle. 1975. Short-fed, grass-fed and long-fed beef compared. *Kansas Agr. Exp. Sta. Rep.* 230.

Lorenzen, C.L., R.K. Miller, J.F. Taylor, T.R. Neely, J.D. Tatum, J.W. Wise, M.J. Buyck, J.O. Reagan, and J.W. Savell. 2003. Beef customer satisfaction: Trained sensory panel ratings and Warner-Bratzler shear force values. *J. Anim. Sci.* 81: 143-149.

Luckett, R.L., T.D. Bidner, E.A. Icaza, and J.W. Turner. 1975. Tenderness studies in straightbred and crossbred steers. *J. Anim. Sci.* 40: 468-475.

Maas, M.R., K.A. Glass, and M.P. Doyle. 1989. Sodium lactate delays toxin production by *Clostridium botulinum* in cook-in-bag turkey products. *Appl. and Environ. Micro.* 55: 2226.

Maca, J.V., R.K. Miller, and G.R. Acuff. 1997. Microbiological, sensory and chemical characteristics of vacuum-packaged ground beef patties treated with salts of organic acids. *J. Food Sci.* 62: 591.

Maca, J.V., R.K. Miller, M.E. Bigner, L.M. Lucia, and G.R. Acuff. 1999. Sodium lactate and storage temperature effects on shelf life of vacuum packaged beef top rounds. *Meat Sci.* 53: 23.

Maca, J.V., R.K. Miller, J.D. Maca, and G.R. Acuff. 1997. Microbiological, sensory and chemical characteristics of vacuum-packaged cooked beef top rounds treated with sodium lactate and sodium propionate. *J. Food Sci.* 62: 586.

Marsh, B.B., T.P. Ringkob, R.L. Russell, D.R. Swartz, and L.A. Pagel. 1987. Effects of early-postmortem glycolytic rate on beef tenderness. *Meat Sci.* 21: 241.

May, S.G., W.L. Mies, J.W. Edwards, F.L. Williams, J.W. Wise, J.B. Morgan, J.W. Savell, and H.R. Cross. 1992. Beef carcass composition of slaughter cattle differing in frame size, muscle score, and external fatness. *J. Anim. Sci.* 70: 2431-2445.

McKeith, F.K., D.O.L. DeVol, R.S. Miles, P.J. Bechtel, and T.R. Carr. 1985. Chemical and sensory properties of thirteen major beef muscles. *J. Food Sci.* 50: 869.

Meilgaard, M., G.V. Civille, and B.T. Carr. 1999. *Sensory Evaluation Techniques*, 3rd ed. CRC Press Inc., Boca Raton, FL.

Melton, S.L. 1983. Effect of forage feeding on beef flavor. *Food Technol.* 37(5): 239-248.

Meyer, B., J. Thomas, R. Buckley, and J.W. Cole. 1960. The quality of grain-finished and grass-finished beef as affected by ripening. *Food Technol.* 14: 4-7.

Miller, R.K., H.R. Cross, J.D. Crouse, and J.D. Tatum. 1987. The influence of diet and time of feed on carcass traits and quality. *Meat Sci.* 19: 303-313.

Miller, R.K., L.C. Rockwell, D.K. Lunt, and G.E. Carstens. 1997. Determination of the flavor attributes of cooked beef from steers fed corn or barley based diets. *Meat Sci.* 44: 235-243.

Mitlohner, F.M., M.L. Gaylean, and J.J. McGlone. 2002. Shade effects on performance, carcass traits, physiology, and behavior of heat-stressed feedlot heifers. *J. Anim. Sci.* 80: 2043-2050.

Mitlohner, F.M., J.L. Morrow, J.W. Dailey, S.C. Wilson, M.L. Gaylean, M.F. Miller, and J.J. McGlone. 2001. Shade and water misting effects on behavior, physiology, performance, and carcass traits of heat stressed feedlot cattle. *J. Anim. Sci.* 79: 2327-2335.

Neely, T.R., C.L. Lorenzen, R.K. Miller, J.D. Tatum, J.W. Wise, J.F. Taylor, M.J. Buyck, J.O. Reagan, and J.W. Savell. 1998. Beef customer satisfaction: Role of cut, USDA quality grade, and city on in-home consumer ratings. *J. Anim. Sci.* 76: 1027-1032.

Offer, G. and J. Trinick. 1983. On the mechanism of water holding in meat: The swelling and shrinking of myofibrils. *Meat Sci.* 8: 245.

Pagach, D.A. 1992. The use of sodium and/or potassium lactate to extend shelf-life and reduce sodium levels in precooked beef systems. MS thesis, Texas A&M University, College Station, TX.

Page, B.T., E. Casas, M.P. Heaton, N.G. Cullen, D.L. Hyndman, C.A. Morris, A.M. Crawford, T.L. Wheeler, M. Koohmaraie, J.W. Keele, and T.P.L. Smith. 2002. Evaluation of single-nucleotide polymorphisms in CAPN1 for association with meat tenderness in cattle. *J. Anim. Sci.* 80: 3077-3085.

Papadopoulos, L.S., R.K. Miller, G.R. Acuff, L.M. Lucia, C. Vanderzant, and H.R. Cross. 1991. Consumer and trained sensory comparisons of cooked beef top rounds treated with sodium lactate. *J. Food Sci.* 56: 1141.

Papadopoulos, L.S., R.K. Miller, L.J. Ringer, and H.R. Cross. 1991. Sodium lactate effect on sensory characteristics, cooked meat color and chemical composition. *J. Food Sci.* 56: 621.

Pike, M.M., T.P. Ringkob, D.D. Beekman, Y.O. Koh, and T. Gerthoff. 1993. Quadratic relationship between early-post-mortem glycolytic rate and beef tenderness. *Meat Sci.* 34: 13-16.

Ramsey, C.B., J.W. Cole, B.H. Meyer, and R.S. Temple. 1963. Effects of type and breed of British, Zebu and dairy cattle on production, palatability and composition. II. Palatability differences and cooking losses as determined by laboratory and family panels. *J. Anim. Sci.* 22: 1001-1008.

Riley, R.R., G.C. Smith, H.R. Cross, J.W. Savell, C.R. Long, and T.R. Cartwright. 1986. Chronological age and breed-type effects on carcass characteristics and palatability of bull beef. *Meat Sci.* 17: 187-198.

Savell, J.W., G.C. Smith, and Z.L. Carpenter. 1977. Blade tenderization of four muscles from three weight-grade groups of beef. *J. Food Sci.* 42: 866-870, 874.

Schroeder, J.W., D.A. Cramer, R.A. Bowling, and C.W. Cook. 1980. Palatability, shelflife and chemical differences between forage- and grain-finished beef. *J. Anim. Sci.* 50: 852-859.

Shackelford, S.D. 1992. Heritabilities and phenotypic and genetic correlations for bovine postrigor calpastatin activity and measures of meat tenderness and muscle growth. PhD Dissertation, Texas A&M University, College Station, TX.

Shackelford, S.D., M. Koohmaraie, M.F. Miller, J.D. Crouse, and J.O. Reagan. 1991. An evaluation of tenderness of the longissimus muscle of Angus by Hereford versus Brahman crossbred heifers. *J. Anim. Sci.* 69: 171-177.

Shackelford, S.D., M. Koohmaraie, and J.W. Savell. 1994. Evaluation of Longissimus dorsi muscle pH at three hours post mortem as a predictor of beef tenderness. *Meat Sci.* 37: 195-204.

Skelley, G.C., R.L. Edward, F.B. Wardlaw, and A.K. Torrence. 1978. Selected high forage rations and their relationship to beef quality, fatty acids and amino acids. *J. Anim. Sci.* 47: 1102-1108.

Smulders, F.J.M., B.B. Marsh, D.R. Swartz, R.L. Russell, and M.E. Hoenecke. 1990. Beef tenderness and sarcomere length. *Meat Sci.* 28: 349.

Tatum, J.D. 1981. Is tenderness nutritionally controlled? *Proc. Recip. Meat Conf.* 34: 65-67.

Tatum, J.D., G.C. Smith, and Z.L. Carpenter. 1978. Blade tenderization of steer, cow and bull beef. *J. Food Sci.* 43: 819-822.

Trout, G.R. and G.R. Schmidt. 1986. Effect of phosphates on the functional properties of restructured beef rolls: The role of pH, ionic strength, and phosphate type. *J. Food Sci.* 51: 1416.

USDA. 1990. Composition of foods: Beef products raw, processed, prepared. Human Nutrition Information Service, Agriculture Handbook 8-13, USDA. Washington, DC.

Vann, R.C., J.C. Paschal, and R.D. Randel. 2004. Relationships between measures of temperament and carcass traits in feedlot steers. *J. Anim. Sci.* 82 (1): 259.

Voisinet, B.D., T. Grandin, S.F. O'Connor, J.D. Tatum, and M.J. Deesing. 1997. *Bos indicus* cross feedlot cattle with excitable temperaments have tougher meat and a higher incidence of borderline dark cutters. *Meat Sci.* 46: 367-377.

Vote, D.J., W.J. Platter, J.D. Tatum, G.R. Schmidt, K.E. Belk, G.C. Smith, and N.C. Speer. 2000. Injection of beef strip loins with solutions containing sodium tripolyphosphate, sodium lactate and sodium chloride to enhance palatability. *J. Anim. Sci.* 78: 952.

Weber, A.J. 1997. Palatability of roast beef and turkey injected with salts of various organic acids. MS thesis, Texas A&M University, College Station, TX.

Wheeler, T.L., M. Koohmaraie, and J.D. Crouse. 1991. Effects of calcium chloride injection and hot boning on the tenderness of round muscles. *J. Anim. Sci.* 69: 4871-4875.

———. 1992. The effect of postmortem time of injection and freezing on the effectiveness of calcium chloride for improving beef tenderness. *J. Anim. Sci.* 70: 3451-3457.

Wheeler, T.L., M. Koohmaraie, J.L. Lansdell, G.R. Siragusa, and M.F. Miller. 1993. Effect of postmortem injection time, injection level, and concentration of calcium chloride on beef quality traits. *J. Anim. Sci.* 71: 2965-2974.

Wheeler, T.L., M. Koohmaraie, and S.D. Shackelford. 1994. Reducing inconsistent beef tenderness with calcium-activated tenderization. *Proc. Meat Ind. Res. Conf.* 119-130.

Wheeler, T.L., J.W. Savell, H.R. Cross, D.K. Lunt, and S.B. Smith. 1990. Mechanisms associated with the variation in tenderness of meat from Brahman and Hereford cattle. *J. Anim. Sci.* 68: 4206-4220.

Wheeler, T.L., M. Koohmaraie, and S.D. Shackelford. 1997. Effect of postmortem injection time and postinjection aging time on the calcium-activated tenderization process in beef. *J. Anim. Sci.* 75(10): 2652-2660.

Whipple, G., M. Koohmaraie, M.E. Dikeman, J.D. Crouse, M.C. Hunt, and R.D. Klemm. 1990. Evaluation of attributes that affect longissimus muscle tenderness in *Bos taurus* and *Bos indicus* cattle. *J. Anim. Sci.* 68: 2716-2728.

Williams, C.B., J.W. Keele, and D.R. Waldo. 1992. A computer model to predict empty body weight in cattle from diet and animal characteristics. *J. Anim. Sci.* 70: 3215-3222.

Wu, J.J., D.M. Allen, M.C. Hunt, C.L. Kastner, and D.H. Kropf. 1981. Nutritional effects on beef palatability and collagen characteristics. *J. Anim. Sci.* 51 (1): 71.

Chapter 8

Beef Quality, Beef Demand, and Consumer Preferences

Wendy J. Umberger

INTRODUCTION

U.S. beef demand declined for nearly 20 years from 1979 to 1998. During this time the industry continuously lost consumer market share to poultry and substitute protein products. After realizing the important relationship between demand and producer profitability, the U.S. beef industry invested substantial research dollars in an effort to address and to understand the problems at the root of their market loss issue. Through a series of beef quality audits, the industry began to realize that in order to regain market share, it needed to focus on improving the relative value of its beef products by focusing on increasing efficiency, quality, and safety throughout the beef supply chain (Purcell, 2002; Smith et al., 1995). It became evident to the beef industry that the final consumer of its product was becoming more and more of a driver of beef demand, and changes affecting the overall U.S. economy were influencing consumers' lifestyles, and thus, their demand for meat products.

Significant efforts were directed toward understanding the end consumers of beef and the factors influencing their meat-purchasing decisions and overall meat demand. Numerous surveys discovered that in addition to having interest in the nutrition and convenience of their meat products, consumers were becoming increasingly concerned

Handbook of Beef Safety and Quality

doi:10.1300/5640_08

about the quality of the beef they purchased. The industry discovered that during the 1980s and 1990s, there was a growing divergence between what the consumer wanted in terms of quality, and what the beef industry offered (Purcell, 2002). In order to regain market share, and to increase market demand, consumer quality issues had to be addressed.

This chapter summarizes what has been learned with respect to the meat-quality factors affecting consumer demand for beef. In order for readers to completely understand the role consumers and their perceptions of quality play in influencing demand, this chapter first provides a definition of demand, and summarizes the determinants of beef demand. This is followed by a discussion of the current economic trends affecting consumer preferences and purchasing decisions. Various definitions of food quality are introduced, and a model, the Total Food Quality Model, is presented to provide an overview of how consumers develop their perceptions of meat quality. Finally, research on consumer preferences and their willingness to pay for various quality attributes is discussed in order to provide ideas to increase demand by developing future branded products that meet the needs of specific market segments.

DEMAND AND CONSUMER BEEF DEMAND DETERMINANTS

Consumers and their perceptions of beef quality and safety have an impact on beef demand (Marsh, Schroeder, and Mintert, 2004; Mintert, Schroeder, and Marsh, 2002; Piggot and Marsh, 2004; Purcell, 1998, 2002; Schroeder, Marsh, and Mintert, 2000). Unfortunately, the term demand is often misused and misunderstood. Demand is not the per capita consumption of a product, nor is it a product's price. Simply stated, demand is what people are *willing and able* to purchase at various prices. Demand is usually illustrated as a curve or a schedule of price and quantity relationships representing the various quantities of goods or services that a consumer or group of consumers (a market) will buy at alternative price ranges. Consumer demand schedules are usually graphed with the quantities (units) sold or consumed of the product on the horizontal axis and the price per unit sold or consumed on the vertical axis of the graph. For example, a graph illustrating the demand curve for ribeye steaks in the United States over a given period of time would have on its horizontal axis the U.S. per capita consumption of ribeye steaks in retail pounds, and on its vertical axis there would be a range of retail prices over which consumers purchased various quantities of the beef products during the time period. (See Mintert, Schroeder, and Marsh, 2002; Purcell, 1998; and Schroeder, Marsh, and Mintert, 2000, for graphical examples of beef demand.)

When economists say that demand for a commodity or product changes, they are referring to shifts in the demand schedule or curve. Shifts to the right (out) of a given demand curve are referred to as increases in demand, and shifts of the curve to the left (in) are decreases in demand. We consider shifts out (in) as increases (decreases), because at any given price, the quantity demanded by consumers will be greater (less) than it was prior to the shift. Demand increases when one of the following occurs: consumers are willing to purchase a larger quantity of the product at the same price, consumers are willing to purchase the same quantity of product at a higher price, or consumers are willing to purchase a higher quantity of the product at a higher price. Assuming other economic factors stay constant (e.g., supply), industry revenues will tend to increase when demand increases (Purcell, 2001).

Quantity demanded is different from demand; it is the specific amount of a good or service that consumers will purchase at a given price. Changes in *quantity demanded* are usually caused by changes in price, holding all other economic factors constant. Changes or shifts in demand result from changes in economic factors other than a good's own price. For the beef industry these demand shifters include factors such as prices of competing products (such as poultry or pork), changing consumer tastes and preferences (possibly due to food safety or health and nutritional concerns), and changing demographics (e.g., income or age distribution) (Mintert, Schroeder, and Marsh, 2002; Schroeder, Marsh, and Mintert, 2000).

Demand indices are also often used to show changes in demand. Purcell (2002) developed a demand index, which is widely used by the beef industry to show the percent change in beef prices if demand had been at a 1980 level. Figure 8.1 is an illustration of this index, and shows a decline in retail USDA Choice grade beef demand from 1980 to 1998. While beef demand began to recover in 1999, beef prices in

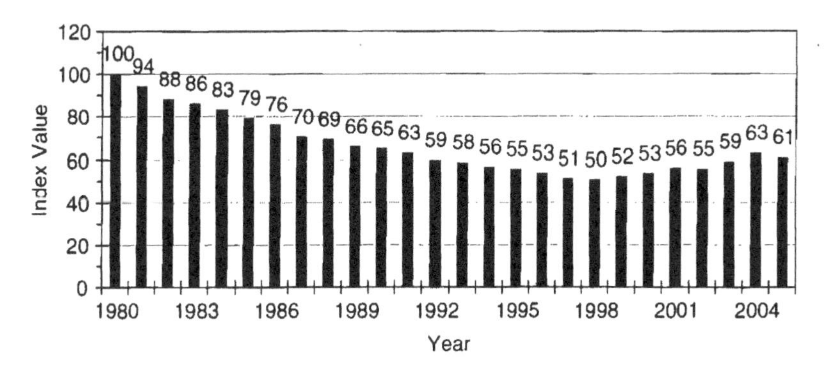

FIGURE 8.1. Annual Retail Choice Beef Demand Index (1980-2005) with price deflated using CPI, 1980 = 100. *Source:* Livestock Marketing Information Center, 2005.

2005 were still estimated to be 39 percent lower than if beef demand had been at the same level as in 1980 (Purcell, 2002; Mintert, Schroeder, and Marsh, 2002). Therefore, while demand has tended to increase over the past seven years, there is still room for improvement.

CONSUMER BEEF-DEMAND MODELS

Numerous researchers have examined beef demand and have developed and estimated demand models in an attempt to represent the factors that explain beef demand (Capps, Moen, and Branson, 1988; Capps and Schmitz, 1991; Kinnucan et al., 1997; Marsh, Schroeder, and Mintert, 2004; McGuirk et al., 1995; Piggott and Marsh, 2004; Schroeder, Marsh, and Mintert, 2000). Initially beef demand was modeled as a function of beef price (own price), prices of competing meats, total meat expenditures, and consumer income. More recent research has suggested that a structural change occurred in meat demand throughout the 1960s and 1980s and that researchers need to consider additional factors such as health information, food safety concerns, and changing market demographics such as ethnicity and the changing labor force in their models. (See McGuirk et al., 1995; Marsh, Schroeder, and Mintert, 2004; Schroeder, Marsh, and Mintert, 2000; and Piggot and Marsh, 2004, for a discussion of numerous previous beef demand studies.) A Beef Demand Study Group Report, published in 1999, identified five important beef demand drivers: food safety, palatability, health and nutrition, consumer-friendly products (convenience), and cost/price/perceived value relative to competing meats.

Schroeder, Marsh, and Mintert (2000) developed a meat demand system to better understand why beef demand declined during the 1980s and most of the 1990s, and to determine how the beef industry could improve demand for beef products. The demand model developed for their study used quarterly data from 1982 to 1998 to estimate the effect of prices of competing meats (poultry and pork), total consumer expenditures on all goods, changing consumer demographics, food safety concerns, and health information on beef demand. The results of the Schroeder, Marsh, and Mintert (2000) study and several other studies provide insight on the role that changing consumers and their preferences for quality and safety have had and will likely continue to have in influencing beef demand. Each of these demand factors will be further discussed in the following paragraphs.

Relative Meat Prices and Consumer Expenditures

As mentioned previously, demand is impacted by both consumers' desire and ability to purchase a product. Prices of protein products influence both a consumer's desire and ability to purchase beef. Consumers will not purchase a good if it is too expensive for their budget, or if it is perceived to be a poor value for the money relative to another available substitute good. All of the numerous factors affecting consumers' perceptions of product value are certainly correlated to their attitudes about quality; in other words, perceived value is a function of quality. Quality as it relates to value will be discussed in subsequent sections of this chapter; however, it is important to point out now that a consumer's perception of a product's quality relative to other substitute products' perceived quality and relevant prices does influence the value a consumer places on a good. For example, when a consumer walks up to a supermarket meat case or sits down at a restaurant and begins to make a decision on what to purchase or what to order, he or she considers the price as well as the beef product's other attributes (e.g., expected taste, nutritional content) and compares them to the price and attributes of the other available substitute protein products. A consumer may want to purchase a steak and may perceive it to be a good value, but he or she may financially only be able to purchase ground beef or a chicken breast. Value, price, and quality are all related and all play a role in beef demand and in the consumer-purchasing decision.

How much does the price of beef versus other substitute goods influence beef demand? Schroeder, Marsh, and Mintert (2000) found poultry and pork to be weak substitutes for beef, meaning that while the relative prices of competing meats matter, per capita beef consumption is only slightly influenced by changes in the prices of pork and poultry. Furthermore, beef demand is fairly inelastic, implying that the quantity of beef demanded changed by less than 1 percent when beef prices increased by 1 percent.[1] Schroeder, Marsh, and Mintert (2000) state that because per capita beef consumption is not highly responsive to its own price or to changes in pork and poultry prices, as consumer incomes rise, and as beef expenditures represent increasingly smaller proportions of consumers' total expenditures, beef demand will become progressively less responsive to changes in prices. This finding "indicates that many consumers are willing to pay for a high quality product (i.e., price is less of an issue if quality is high). As a result, consideration should be given to devoting resources to research focusing on quality . . ." (Schroeder, Marsh, and Mintert, 2000, p. 2).

Although beef demand appears not to be highly sensitive to relative price changes, it is very responsive to changes in total consumer expenditures. Estimates from the model indicate that beef demand increased 0.90 percent for every 1 percent increase in total per capita disposable expenditures. Therefore, when incomes and spending increase, beef demand tends to grow, but decreasing disposable incomes or increases in savings can have a negative effect on beef demand (Schroeder, Marsh, and Mintert, 2000).

Food Safety and Health Concerns

In addition to Schroeder, Marsh, and Mintert (2000), other researchers have examined the impact of food safety and health concerns on consumer demand. As noted by Piggott and Marsh (2004), it is important to differentiate between the impacts of food safety "outbreaks" and health or nutritional concerns. Illnesses caused by food safety "outbreaks" can result in short-run, long-run, and possibly even fatal illnesses. "Outbreaks are characterized by unanticipated and sudden food safety events that shock demand and are followed by an outpouring of public information" (Piggott and Marsh, 2004, p. 158). Alternatively, the impacts of health-related attributes of concern such as cholesterol cannot always be determined immediately and usually require longer periods of continuous consumption to be detected and may consequently have a longer-run impact on demand (Piggott and Marsh, 2004).

Several studies have examined the effects of food safety outbreaks and product recalls by using developed food safety indices in their demand models. Piggott and Marsh's (2004) food safety index was based on the number of media reports (newspaper articles) published regarding food safety events (e.g., product recalls due to *E. coli* O157:H7 and/or issues related to BSE) related to meat products. From 1982 to 1999, beef versus pork or poultry had the highest quarterly average number of articles related to food safety issues (a mean of 174.2 per quarter for beef versus 143.1 and 153.0 for pork and poultry, respectively). The average number of food safety articles related to only beef products increased 29 percent from the 1982-1989 to the 1990-1999 period; this was due primarily to an increase in the number of BSE articles (Piggott and Marsh, 2004). Using the media index as a measure, the average response of beef demand to food safety concerns was found to be negative and small relative to price and expenditure effects. Larger, short-run, negative impacts occurred when prominent food safety events occurred.

Food safety impacts on demand were also evaluated using the number of USDA, FSIS product recalls of beef products for the period of 1982 to 1998 (Marsh, Schroeder, and Mintert, 2004; Schroeder, Marsh, and Mintert, 2000). During this period, recalls for all meat products trended upward; however, beef had more recalls on average than pork or poultry (2.20, 2.02, and 1.56 recalls, on average, per quarter for beef, pork, and poultry, respectively). For beef products, this higher level of product recalls may be due to the increased attention and increased ability to detect *E. coli* O157:H7 in the supply chain (Marsh, Schroeder, and Mintert, 2004; Schroeder, Marsh, and

Mintert, 2000). The results of the demand analyses using product recall indices as a measure of food safety were similar to those found by Piggott and Marsh (2004) when using a media index. Beef product recalls by the USDA, FSIS were found to have a significant negative effect on beef demand, but again, similar to media reports, product recalls only explained a small portion of consumer demand, when compared to the effects of prices and income (Marsh, Schroeder, and Mintert, 2004; Schroeder, Marsh, and Mintert, 2000). Nonetheless, while the impacts were small due to a generally small number of product recalls, the results suggest that beef demand would see a significant decline if a large increase in beef recalls occurred. This outcome implies that the beef industry must maintain a proactive food-safety approach in order to continue increasing consumer demand for beef (Marsh, Schroeder, and Mintert, 2004; Schroeder, Marsh, and Mintert, 2000).

In the late 1970s and 1980s, U.S. consumers started to be more concerned about their diet and health, and a large focus was put on the fat content and cholesterol levels in meats (Purcell, 2002). The effects on beef demand of health information or consumer attitudes toward fat have been modeled by several researchers (Capps, Moen, and Branson, 1988; Kinnucan et al., 1997; McGuirk et al., 1995; Schroeder, Marsh, and Mintert, 2000). The results obtained by Kinnucan et al. (1997), McGuirk et al. (1995), and Schroeder, Marsh, and Mintert (2000) are similar; as consumers were exposed to more information supporting the linkage between cholesterol and heart disease, beef demand weakened while poultry and pork demand increased. Schroeder, Marsh, and Mintert (2000) emphasize the need for continued research to better understand the nutritional properties of beef and increased consumer education to convey truthful information with respect to the dietary benefits of consuming beef.

Changing Consumer Demographics

Not surprisingly, changing consumer demographics such as average age, income, and ethnicity, have had an impact on U.S. consumer lifestyles and as a result on their demand for meat products. The growing number of females in the workforce and homes with two wage earners increased the need for convenient food products (Farm

Foundation, 2006; McGuirk et al., 1995; Schroeder, Marsh, and Mintert, 2000). Results from the Schroeder, Marsh, and Mintert (2000) model indicate that the increasing percentage of females in the workforce from 1982 to 1998 had a significant and negative impact on beef and pork consumption, but a positive impact on poultry consumption. For each 1 percent increase of females in the labor force, per capita beef consumption declined by 1.51 percent; however, poultry consumption increased by 0.46 percent. Until recently, the beef industry lagged behind the poultry industry in development of convenience product lines. The lack of available convenience beef products may be one reason beef demand declined during this period. Convenience is obviously an important product attribute for consumers.

Increasing incomes indirectly affect beef demand when total consumer expenditures simultaneously increase (as discussed previously). Consumer-income levels also impact the types of products consumers demand; thus, consumer income will likely play a greater role in influencing beef demand in the future. As consumers' disposable incomes grow, they are able to move beyond just being concerned about consuming food that meets their perceived basic nutritional needs. Consumers may also begin to be concerned about the impact that individual food production decisions have on other people, the environment, and animals (Tronstad et al., 2005). The consumer food demand pyramid (Figure 8.2) illustrates the role that income plays in the consumer choice process (Kinsey, 2000). The food demand pyramid suggests that at lower income levels, consumers focus

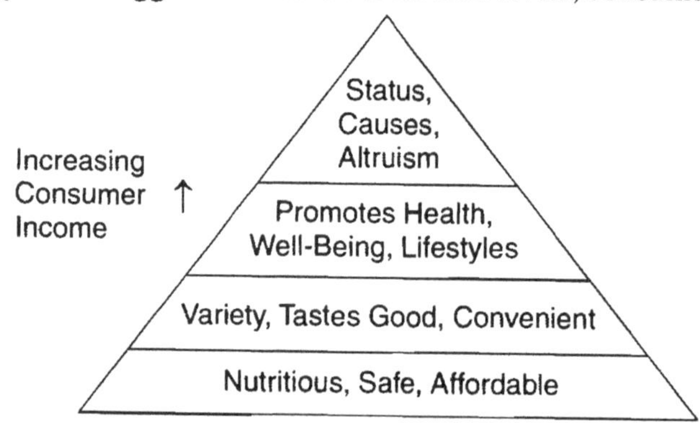

FIGURE 8.2. Consumer food demand pyramid. *Source:* Adapted from Kinsey (2000).

first on the base of the pyramid, obtaining affordable and safe food products that provide them with sufficient calories and nutrients. Once basic nutritional needs are met, consumers then become interested in food attributes such as taste and convenience, which are further up the pyramid. Consumers at higher income levels want expanded information about their food, and knowledge about how food production affects their health and lifestyle (Tronstad et al., 2005). They may use their money to purchase food products that support the various causes they are interested in, based on their values and beliefs. As incomes increase, the demand for food products with different characteristics evolves, consumers' definition of quality changes, and segments of the market begin to demand beef with additional attributes.

WHAT IS BEEF QUALITY AND HOW DO CONSUMERS DEVELOP THEIR PERCEPTIONS OF QUALITY?

It is apparent from the demand models discussed in the previous section that beef demand is a function of quality. Twenty years ago, a safe and high-quality beef product may have largely meant the same thing to most beef consumers. Historically, the primary method used to communicate beef quality to consumers was the USDA beef quality grading system.[2] While consumers often had information on the USDA quality grade of a beef product, very few consumers knew how to interpret quality grades (Killinger et al., 2004; Umberger, 2001). In fact, several studies found the USDA quality-grading system to be inefficient due to discrepancies between the USDA grades and consumers' visual and taste preferences for beef, which were driven by health concerns regarding fat and cholesterol or lack of knowledge about the attributes of beef that contribute to the eating experience (Acebron and Dopico, 2000; Cox, McMullen, and Garrod, 1990; Lusk, 2001; Steenkamp and Van Trijp, 1996; Umberger, 2001). Therefore, because beef products sold at supermarkets traditionally have been relatively undifferentiated and unbranded, consumers' quality perceptions regarding beef products were established based on past eating experiences and the quality and safety of the previous products they consumed.

Due to confusion regarding the information disseminated in USDA quality grades, the industry realized that the grading system may be of limited use to consumers when determining quality and choosing beef products (Purcell, 2001, 2002). Over time, numerous branded meat products labeled with attributes such as natural, lean, organic, free range, and grass-fed have been developed to meet heterogeneous consumer demands for quality. The plethora of new information constantly reaching consumers through new product introductions, government reports, the media, and other outlets have all caused consumers' perceptions of beef quality and safety to evolve. Today, more than ever, consumers use multiple attributes to evaluate the quality of their food products and to determine their preferences for one meat product over another.

Even now, quality is described rather ambiguously and quality still means different things to different people and industries. The International Organization of Standardization's (ISO, 1994) definition of food quality is probably the most widely agreed upon definition of quality across disciplines. The ISO defines quality as "the totality of characteristics of an entity that bear on its ability to satisfy stated and implied needs." Again, this is a very broad definition, as it basically says quality is determined by all of the features a consumer wants, yet very few consumers desire the same product features. When discussing beef products, multiple definitions of beef quality exist. These definitions not only differ across academic disciplines, but individual consumers also have different ideas of what defines quality red meat. How exactly does a consumer define quality when it comes to beef? One consumer may define a safe, quality beef product as one that is USDA-inspected, is highly marbled, and is very tender, and another consumer may see a safe, quality beef product as one that is lean and was produced via natural production methods.

Product Quality Characteristics, Attributes, and Cues

Becker (2000) discusses the evolution of the term quality over time and suggests that the food-quality approach historically used by natural scientists, such as food scientists and meat scientists, is one based primarily on measuring quality characteristics defined as *product characteristics*. "Product characteristics are those features of a product which are used as (technical indicators) for product quality and are (in principle) measurable with standardized analytical (including sensoric) methods" (Becker, 2000, p. 163). The food science literature on meat quality establishes four categories of technical product characteristics relevant to meat quality: (1) nutritional value (e.g., protein, fat, or carbohydrate content); (2) processing quality (e.g., shear force, color, pH value, water-binding capacity); (3) hygienic-toxicological quality (e.g., amount of residues, microbial status, additives); and (4) sensory quality (e.g., tenderness, juiciness, flavor, marbling) (Becker, 2000). While these objective product characteristics are important to consumers, this traditional, natural science approach does not consider the many other factors influencing consumers' perceptions of quality in today's marketplace.

Rather than using the product characteristics approach that focuses on measuring quality through technical indicators, the quality approach taken in the consumer behavior literature focuses on product attributes and the consumer's subjective perceptions of these attributes. Becker (2000) makes a clear distinction between technical product characteristics and product *attributes*. He defines product attributes as "those features of a product meeting consumer needs" (p. 163). Product attributes include the technical product characteristics, but are broader and include information on other product characteristics that consumers perceive to affect quality and subsequently value.

Product attributes are generally categorized as *search, experience,* or *credence attributes* (Caswell, 1998; Mojduszka, 1996; Darby and Karni, 1973; Nelson, 1970). Search attributes can be evaluated by consumers at the point of purchase and prior to consumption. Examples of search attributes in meat are color, leanness, marbling (intramuscular fat), brand name, food safety inspection, and price. Experience attributes are only observable during or following consumption and include the safety and eating quality (e.g., tenderness, juiciness, flavor, and smell) of a meat product. Credence attributes may be perceived by the consumer to be important, but he or she cannot discern their presence when purchasing a product, or even after normal use. The origin and production practices used to produce the meat (e.g., organic, natural, hormones, antibiotics) are examples of credence attributes. These categories of attributes can also be classified in terms of *process* and *product* attributes. Process attributes in meat include the production methods used to raise the meat animal: antibiotics, hormones, organic, and animal welfare to name a few. Product attributes, on the other hand, include many of the technical characteristics of food products, and are related to the product's food safety, nutritional content, sensory and functional characteristics, and image (Caswell, Bredahl, and Hooker, 1998; Grunert, 1997; Northern, 2000).

Consumers develop information about a product's various attributes by using *cues* they receive when visually inspecting products while shopping and through consuming products (Becker, 2000; Northern, 2000). Quality cues are used by consumers to develop their perceptions of quality prior to purchase (Northern, 2000; Oude Ophuis and Van Trijp, 1995). Product-quality attribute cues are different from the product-quality information consumers receive from the media or through word-of-mouth (Becker, 2000).

Product-quality cues are typically further categorized as *intrinsic* or *extrinsic cues*. Intrinsic cues are attributes inherent to the product that cannot be changed without changing the physical properties of the product. Color, smell, leanness, and marbling are examples of intrinsic cues for beef steaks. Extrinsic cues are defined as attributes that are only related to the physical product; they are unique because they can be manipulated without changing the physical product (Caswell, Bredahl, and Hooker, 1998; Northern, 2000; Oude Ophuis and Van Trijp, 1995). Informational food labels, certifications, and brands are commonly used as extrinsic quality cues and are the only way to indicate the presence of certain credence process attributes related to production processes such as organic and/or natural, antibiotic and/or hormone use, animal welfare, and traceability. Extrinsic cues may also be used to indicate credence product attributes related to food safety, nutrition, and even texture (e.g., tenderness verification labeling) (Becker, 2000; Caswell, Bredahl, and Hooker, 1998; Grunert, 1997; Nelson, 1970).

Intrinsic and Extrinsic Cues' Role in Forming Consumer Quality Perceptions

Northern (2000) outlines important differences between intrinsic and extrinsic cues and their role and ability to signal quality information to consumers. Intrinsic cues are only able to provide consumers with information about search and experience attributes. In order for intrinsic cues to effectively communicate quality to consumers, consumers need help understanding the role that intrinsic cues such as beef cut, color, and marbling play in determining the eating experience and quality of beef. For example, consumers may fear fat due to nutritional concerns, and may not understand the importance of intramuscular fat (marbling) in determining eating quality. Consumers who perceive marbling to be a negative quality cue because of nutritional concerns and who purchase low marbled meat may experience an unpalatable eating experience and subsequently, after numerous bad experiences, may shift their meat consumption away from beef altogether (Killinger et al., 2004; Umberger, 2001). Northern (2000) also indicates the importance of voluntary and mandatory industry-wide technical standards for maintaining quality-management systems. The Beef Quality Assurance program and HAACP standards are examples of quality-management systems and technical standards that are currently used in U.S. beef supply chains.

Extrinsic cues can signal both experience and credence attributes to consumers. However, extrinsic cues are only effective when the cue is determined by a credible source (Northern, 2000). Credibility is often verified through certification by a reputable third party. For meat products, consumer studies have found that U.S. consumers trust and prefer the government over other certification agencies; however, other reputable third-party agencies include nongovernment organizations (NGOs), religious organizations, and private companies (Christensen, 2002; Christensen et al., 2003; Loureiro and Umberger, 2005; Wessells, Johnston, and Donath, 1999). Certification by third-party certifying agencies provides increased credibility through production and quality standards, which must be met before the product can carry the certified label. In addition, third-party certifying agencies perform product testing, certification of producers and processors, and enforcement of standards.

The USDA branches of the Food and Drug Administration (USDA, FDA), the Food Safety Inspection Service (USDA, FSIS), and the Agricultural Marketing Service (USDA, AMS) are all reputable and trusted third-party agencies that verify extrinsic quality attributes in beef. An example of third-party standards overseen by the government are the USDA organic and natural standards. Consumers purchasing beef that is labeled as "USDA Certified Organic" can look up the specific set of practices that must be followed in order for the beef to be certified as "organic." The USDA also authorizes private and not-for-profit organizations to certify organic food in the United States. The USDA then requires accredited certifying organizations to maintain and submit records to the USDA annually. Another example of a government third-party certification program is the Process Verified Program (PVP) for meat products, which is overseen by the USDA, AMS. The PVP can be used to provide credibility for labeling product attributes such as the genetic background, age, and source of the animal where the meat came from (e.g., Creekstone Farms and Maverick Natural Meats). Tronstad et al. (2005) provide numerous examples of other available third-party certifications.

Increasing incomes have allowed consumers to move beyond just caring about the taste and nutritional content of their food and to be more concerned about credence attributes. As consumers' value of credence attributes increases, extrinsic cues will play a larger part in influencing consumers' perceptions of meat quality. Therefore, it is important to understand the interrelationship between intrinsic and extrinsic cues and their distinct roles in forming consumers' perceptions of meat quality. The product-attributes approach to food quality is valuable background for understanding the various attributes and cues that might influence consumers' perceptions of meat quality. However, the approach does not explain the process that consumers use when synthesizing the intrinsic and extrinsic cues to develop their quality perceptions and ultimately to make a decision about what meat product to consume.

Grunert (1997) integrated the alternate food-quality approaches and developed a useful conceptual framework for examining how product characteristics (either concrete or abstract) are linked to both the functional and the psychological consequences of consumption. The model, termed the Total Food Quality Model (TFQM), assumes that consumers use observable indicators (such as marbling), termed search characteristics, to form expectations about the experienced quality of a food product. After consuming a product, consumers may compare the experienced quality of the product to the expected quality of the product and adjust the way they make future quality evaluations regarding that product (Grunert et al., 1996).

The TFQM, shown in Figure 8.3, assumes that the expected quality that consumers have for a food product is based on both intrinsic and extrinsic cues. Initially, expected quality and cost cues determine consumers' intention to purchase a product, and ultimately, intrinsic and extrinsic cues, combined with experienced quality, will determine the probability of consumers purchasing similar products in the future. Research on consumers' perceived meat quality using the TFQM as a framework suggests consumers do use a multitude of intrinsic, extrinsic, and cost cues when making meat product-purchasing decisions (Acebron and Dopico, 2000; Bredahl, Grunert, and Fertin, 1998; Poulsen et al., 1996; Steenkamp and Van Trijp, 1996).

Hoffmann (2000) developed a model of food quality similar to the TFQM; however, he distinguished between expected food quality

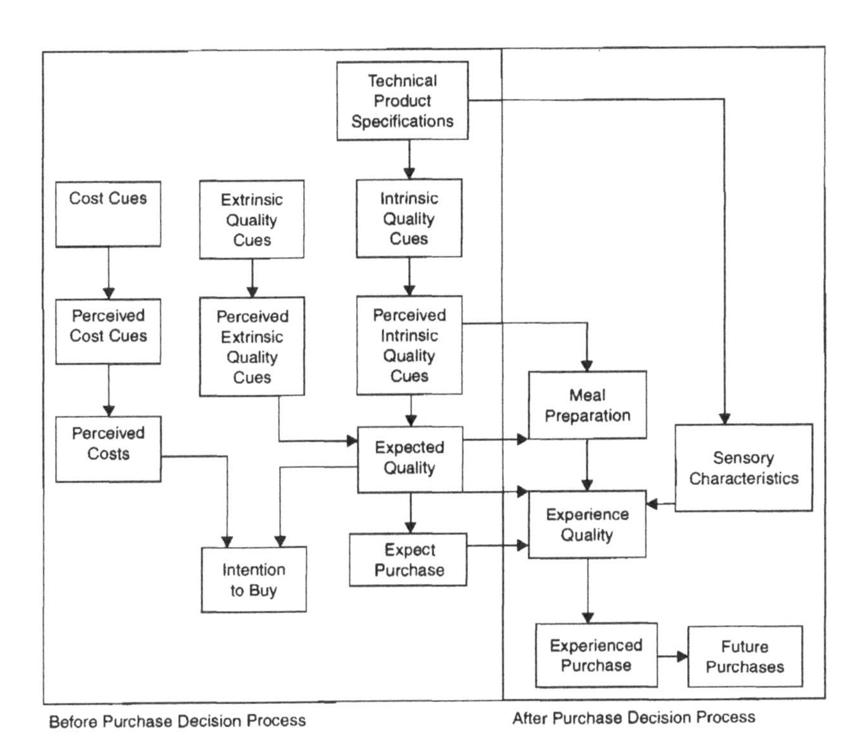

FIGURE 8.3. The Total Food Quality Model. *Source:* Adapted from Grunert et al. (1996).

and food safety. In addition, Hoffmann (2000) suggested that individual consumers only perceive certain quality cues as important and, therefore, he hypothesizes that different consumers use different quality cues when predicting the quality of meat products. Hoffmann (2000) proposed that individual consumer socioeconomic factors, past purchasing behavior, previous experience, attitudes, and preferences determine the quality cues a given consumer considers to be important and consequently will use when purchasing meat. U.S. consumer studies on preferences and willingness to pay for food attributes support Hoffmann's (2000) theory and socioeconomic factors do appear to influence consumers' decision making when selecting food (e.g., Rimal, 2002, 2005; Nayga, 2000). The remainder of this chapter presents an overview of several consumer studies that measure consumer preferences for various intrinsic and extrinsic product cues for beef.

CONSUMER PREFERENCES FOR QUALITY ATTRIBUTES AND CUES

The demand models discussed in the first section of this chapter focused on explaining overall or aggregate market demand for beef; nevertheless, consumer preferences for beef quality attributes are increasingly heterogeneous. Diverse consumer tastes and preferences are driving the formation of multiple market segments and allow for targeted consumer marketing of beef products with attributes that meet or exceed each of the various segment's unique perceptions of high-quality beef. Recent consumer preference studies and research on willingness to pay provide insight on current and future product attributes and cues that are important to segments of consumers, and which may be used to further increase the value of beef products and to grow beef demand.

Beef Attribute Rankings and Ratings

Numerous surveys have explored the relative importance of various attributes in consumers' steak beef-purchasing decisions. Many of these studies have asked consumers to rank or rate the importance of sets of attributes when making beef-purchasing decisions. The outcomes of these studies differ depending on the sets of attributes that consumers were presented with and on whether consumers were asked to rate or rank attributes. For example, Lusk (2001) used responses from a mail survey sent to 650 U.S. consumers to determine the relative importance of six attributes in consumers' steak-purchasing decisions: price, external fat, USDA Quality Grade, brand (label), color, and marbling. On average, consumers ranked brand as least important and ranked color and marbling as most important, followed by USDA quality grade, price, and external fat.

Loureiro and Umberger (2005) also used a U.S. mail survey to determine the desirability of 15 quality cues and attributes that consumers may use when purchasing meat. Consumers in this study were asked to *rate* rather than *rank* the cues and attributes. Food safety inspection and freshness were rated statistically higher on a five-point Likert scale than any other cue or meat attribute included in the set. High-quality grade was rated as "extremely desirable" to "very desirable," and U.S. origin, visual presentation, leanness, tenderness, and nutritional content were ranked as "very desirable" on average. Brands, meat produced or raised locally, and organic production methods were the cues and attributes with the three lowest ratings. These relative ratings were similar to those found in other beef preference studies (see Loureiro and Umberger, 2003; Thilmany, Umberger, and Ziehl, 2006; and Umberger et al., 2003).

The findings of these preference studies further support the notion that various cues and attributes are perceived to be important to consumers when making beef-purchasing decisions. How does the presence of quality attributes and cues such as marbling, tenderness, origin, traceability, and production methods affect the value consumers place on food products?

Marbling

As mentioned previously, marbling has long been the primary determinant of USDA beef-quality grades, yet there appears to be a misunderstanding about how this quality cue is actually used by consumers when USDA quality grades are not labeled. Marbling is an intrinsic quality cue and can be perceived as both desirable and undesirable. Consumer taste tests were combined with experimental auctions to determine the role of marbling in consumer purchasing decisions and willingness to pay. Pairs of steaks that differed in marbling level (Modest/Moderate, USDA Choice versus Slight, USDA Select) were similarly packaged in clear overwrapped styrofoam meat trays, and presented to consumers in Chicago and San Francisco for visual evaluation. Consumers were asked to select their preferred steak for each pair, explain their selection criteria, and provide a bid indicating their willingness to pay for each steak in the pair. The majority of the consumers, 68.7 percent, visually preferred and were willing to pay an average of $1.26 more per pound for the lower-marbled, USDA Select steak. Approximately 20 percent of the consumers visually preferred and were willing to pay an average of $0.89 more per pound for the higher marbled, USDA Choice steak; and 12 percent of the consumers were not willing to pay a premium for their preferred steak (Umberger, 2001; Killinger et al., 2004).

Consumers' comments on the steaks that they visually evaluated were classified into five categories: marbling, fat, appearance, color, and overall perceived palatability and cooking quality. The selection criteria used by consumers who preferred the high-marbled steak were different from the criteria used by consumers who chose the low-marbled steak in the visual evaluation. Sixty-five percent of the consumers who preferred the high-marbled steak listed marbling as a selection criterion. Sixty-two percent of the consumers who preferred the low-marbled steak listed fat as a selection criterion. Marbling is an intrinsic cue used by almost all consumers when selecting a steak. However, the majority of the consumers in this study (62 percent) perceived marbling to be a negative attribute related to high fat content more often than a positive factor related to increased palatability (Umberger, 2001; Killinger et al., 2004). These results are similar to those of Acebron and Dopico (2000), where fat was found to have a negative effect on the expected quality of beefsteaks. A model was developed to determine the impact of demographics, consumer selection criteria, and beef knowledge on the probability that a consumer would prefer and be willing to pay a premium for steaks with specific marbling levels. Consumers from San Francisco and those with at least a college education were more likely to prefer the higher-marbled steak (Umberger, 2001).

Tenderness

Some industry experts and academics have suggested the need for tenderness measures to be incorporated into the USDA Quality grading system, due to the important role that tenderness can play in determining palatability and consumer satisfaction (Purcell, 2002; Smith et al., 1995). Tenderness is an experience attribute; therefore, unless it is labeled, it is difficult for the consumer to use tenderness as a quality signal (unless they have the uncommon opportunity to taste the beef product before making a purchasing decision). Tenderness guarantees are now being used by some branded beef companies to signal additional quality to consumers (e.g., Nolan Ryan's All Natural Tender Aged Beef). Research suggests that consumers may value tenderness verification and labeling of tenderness as a quality cue. For example, the majority of participants in studies by Miller et al. (2001), Shackelford et al. (2001), and Lusk and Fox (2002) preferred and were willing to pay more for steaks that were guaranteed tender by the retailer. Consumers in a study conducted by Lusk et al. (2001) were willing to pay an average premium of $1.84 more per pound for a steak labeled "guaranteed tender." In another consumer study, Feldkamp, Schroeder, and Lusk (2005) found that consumers were willing to pay $0.95 more for the guaranteed tender steak. These results indicate that tenderness is an important beef-quality attribute, suggest the future potential of tenderness certifications and labels as value-added attributes, and support the notion of incorporating measures of tenderness into the USDA Quality Grades.

Origin, Traceability, and Production Methods

An area of increasing differentiation among meat products relates to providing consumers with information or labels regarding the source of origin, traceability, methods of production used to raise animals, and processes used to produce meat. Consumers have indicated through surveys and experimental auctions a willingness to pay for meat products labeled with extrinsic cues that verify credence attributes such as "natural," "organic," "traceable-to-the farm," "free-range," "GMO-free," "not produced with hormones and/or antibiotics," "certified humane," "environmentally friendly," and "local" to name a few (Baker and Burnham, 2001; Dickinson and Bailey, 2002;

Loureiro and Umberger, 2003, 2005; Lusk and Fox, 2002; Lusk, Roosen, and Fox, 2003; McGarry-Wolf and Thulin, 2000; Roosen, Lusk, and Fox, 2003; Nayga et al., 2005; Thilmany, Umberger, and Ziehl, 2003, 2006; Umberger et al., 2003; Wirthgen, 2005). These attributes usually involve the use of traceability systems, and require third-party certification in order to verify the integrity of the processes used to produce the products (Tronstad et al., 2005).

The premiums elicited through consumer research often vary across studies for the same attributes. Willingness-to-pay values elicited in consumer studies tends to be hypothetical and is a function of the research method used (e.g., surveys versus experimental auctions), the wording of the questions, and the consumer sample studied. For example, the willingness-to-pay values elicited in three separate Country-of-Origin Labeling (COOL) studies found premiums ranging from as high as 58 percent to approximately 3 percent for "Certified U.S."- or "Guaranteed U.S."- labeled beef. Furthermore, these premium values are usually expressed as sample averages. Average premium values are interesting, in that they indicate to the industry that specific credence attributes may be of value to consumers; however, it is important to realize that these premium values are not necessarily the levels of premiums that would actually exist in the marketplace.

Past research results suggest that the relative importance of these diverse attributes to consumers varies depending upon the demographic characteristics of consumers. For example, studies which found average premiums for COOL also discovered that not all consumers valued COOL as a quality cue, and in fact, some consumers even discounted this type of origin label (Loureiro and Umberger, 2003, 2005; Umberger et al., 2003). Often willingness-to-pay results are more interesting and more useful for developing marketing strategies when demographic factors are included in analysis. For example, Loureiro and Umberger (2003) found that higher-income, female consumers with children in the household were more likely to pay for beef that was labeled as "U.S. Certified" beef. Roosen, Lusk, and Fox (2003) and Lusk, Roosen, and Fox (2003) found regional differences in European and U.S. consumers' relative willingness to pay and preferences for beef attributes.

Traditional economic/demographic factors used to predict market segments of consumers preferring and willing to pay for various credence attributes are not always significant. Sunding (2003) asserts that some consumers may be motivated to purchase and to pay premiums for products labeled with attributes such as "locally raised," "natural," or "certified humane" for altruistic reasons, in addition to incentives related to perceived private benefits. Results from recent research suggest that consumer attitudinal and preference variables such as knowledge, motivation, risk preferences, and environmental and social consciousness play a significant role in determining consumer preferences for credence attributes in beef (Baker and Burnham, 2001; Hoffmann, 2000; Nayga et al., 2005; Thilmany, Umberger, and Ziehl, 2006; Umberger et al., 2003; Wirthgen, 2005). Thilmany, Umberger, and Ziehl (2006) suggest that cues signaling the presence of credence attributes perceived to enhance beef quality will become increasingly important to a growing number of consumers. The beef industry (or segments of the industry) would likely gain by differentiating beef products through the use of labels denoting these credence attributes and developing marketing strategies targeted at consumers who value these attributes.

SUMMARY

Increasing incomes, more women in the workforce, growing ethnic populations, nutritional information, and food safety concerns have all impacted consumers' demand for beef. These factors will continue to play a role in shaping consumers' definition and perceptions of beef quality. Consumer preferences for beef quality attributes are becoming more heterogeneous and the consumer beef market is becoming more segmented. Multiple opportunities exist for niche producers who are interested in target marketing their differentiated beef products. In order to continue the increasing trend in beef demand, the beef supply chain will have to continue to research consumer preferences for beef attributes in an effort to understand their multiple and evolving definitions of beef quality. Beef products marketed with various extrinsic attributes valued by consumers will likely be successful in the short run. However, long-run success requires providing consumers with beef products that not only meet their perceptions of quality based on labeled attributes, but which also provide them with a predictable and high-quality eating experience.

NOTES

1. Schroeder, Marsh, and Mintert (2000) use their model to estimate beef demand elasticities. These elasticity values provide estimates of the percentage change in the quantity demanded of beef given a 1 percent change in beef, poultry, or pork prices. An elasticity with an absolute value of less than one is called inelastic. The Schroeder, Marsh, and Mintert (2000) model's own-price elasticity estimate for beef was –0.61, indicating that from 1982 to 1998, on average, the quantity demanded of beef decreased 0.61 percent when beef price increased by 1 percent. The cross-price beef demand elasticities with respect to pork and poultry prices were 0.04 and 0.02, respectively. Therefore, based on the model results, on average, from 1982 to 1998, beef demand increased by 0.04 and 0.02 percent when pork and chicken prices increased by 1 percent, respectively. Elasticity estimates from other beef-demand studies are presented and discussed in Schroeder, Marsh, and Mintert (2000, pp. 11-12).

2. The USDA quality grades are based largely on the level of marbling or intramuscular fat. As marbling increases, the USDA grades change from Select to Choice to Prime, and the steaks are expected to have a richer flavor, to be more tender, and to be more juicy. USDA Choice and Prime beef products are generally priced at premiums over USDA Select beef products.

LITERATURE CITED

Acebron, L.B. and D.C. Dopico. 2000. The importance of intrinsic and extrinsic cues to expected and experienced quality: An empirical application to beef. *Food Qual. Pref.* 11: 229-238.

Baker, G.A. and T.A. Burnham. 2001. Consumer response to genetically modified foods: Market segment analysis and implications for producers and policy makers. *J. Agric. Res. Econ.* 26 (2): 387-403.

Becker, T. 2000. Consumer perception of fresh meat quality: A framework for analysis. *British Food J.* 102 (3): 158-176.

Beef Demand Study Group Final Report. 1999. Unpublished report of Beef Demand Study Group, ed. P. Genho, November 30.

Bredahl, L., K.G. Grunert, and C. Fertin. 1998. Relating consumer perceptions of pork quality to physical product characteristics. *Food Quality and Preference.* 9: 273-281.

Capps, O. Jr., and J.D. Schmitz. 1991. A recognition of health and nutrition factors in food demand analysis. *W. J. Agric. Econ.* 16 (July): 21-35.

Capps, O. Jr., D.S. Moen, and R.E. Branson. 1988. Consumer characteristics associated with the selection of lean meat products. *Agribusiness* 4: 549-557.

Caswell, J.A. 1998. How labeling of safety and process attributes affects markets for food. *Agric. Res. Econ. Rev.* 27 (October): 151-158.

Caswell, J.A., M.E. Bredahl, and N.E. Hooker. 1998. How quality management systems are affecting the food industry. *Rev. Agric. Econ.* 20: 547-557.

Caswell, J.A. and E.M. Mojduszda. 1996. Using informational labeling to influence the market for quality in food products. *Am. J. Agric. Econ.* 78 (5): 1248-1253.

Christensen, B.J. 2002. Consumer preferences for public and private sector certifications of beef products in the United States and the United Kingdom. International MBA dissertation, Royal Agricultural College, Cirencester, England, and Utah State University, Logan, UT.

Christensen, B.J., D. Bailey, L. Hunnicutt, and R. Ward. 2003. Consumer preferences for public and private sector certifications for beef products in the United States and the United Kingdom. *Intern. Food Agrib. Mgmt. Rev.* 6 (3): 19-39.

Cox, L.J., B.S. McMullen, and P.V. Garrod. 1990. An analysis of the use of grades and housebrand labels in the retail beef market. *W. J. Agric. Econ.* 15 (December): 245-253.

Darby, M.R. and E. Karni. 1973. Free competition and the optimal amount of fraud. *J. Law and Econ.* 16 (1): 67-68.

Dickinson, D.L. and D. Bailey. 2002. Meat traceability: Are U.S. consumers willing to pay for it? *J. of Agric. Res. Econ.* 27: 348-364.

Farm Foundation. 2006. Consumer issues and demand (Chapter 6). In: *The Future of Animal Agriculture in North America.* Primary author, H. Jensen. Publication forthcoming and available online at: www.farmfoundation.org/projects/04-32 FutureofAnimalAg.htm. Accessed on February 22, 2006.

Feldkamp, T.J., T.C. Schroeder, and J.L. Lusk. 2005. Determining consumer valuation of differentiated beef steak quality attributes. *J. of Muscle Foods* 16: 1-15.

Food Safety and Inspection Service, United States Department of Agriculture (USDA, FSIS). 2003. Meat and poultry labeling terms. August. Available online at: www.fsis.usda.gov/oa/pubs/lablterm.htm. Accessed January 17, 2006.

Grunert, K.G. 1997. What's in a steak? A cross cultural study on the quality perception of beef. *Food Qual. and Pref.* 8: 157-174.

Grunert, K.G., H. Larsen, T.K. Madsen, and A. Baadsgaard. 1996. *Market Orientation in Food and Agriculture.* Kluwer Publishing, Boston, MA.

Hoffmann, R. 2000. Country of origin—a consumer perception perspective of fresh meat. *British Food J.* 102 (3): 211-229.

International Organization of Standards (ISO). 1994. Available at: www.iso.org/iso/en/iso9000-14000/understand/basics/general/basics_4.html. Accessed January 17, 2006.

Killinger, K.M., Calkins, C.R., W.J. Umberger, and D.M. Feuz. 2004. Consumer visual preference and value for beef steaks differing in marbling level and color. *J. Anim. Sci.* 82 (November): 3288-3293.

Kinnucan, H.W., H. Xiao, C.J. Hsia, and J.D. Jackson. 1997. Effects of health information and generic advertising on the U.S. meat industry. *Am. J. Agric. Econ.* 79: 13-23.

Kinsey, J. 2000. The changing global consumer. Presented at the 2000 IAMA World Food and Agribusiness Congress. Chicago, IL.

Livestock Marketing Information Center (LMIC). 2005. Lakewood, CO. Available at: www.lmic.info. Accessed January 17, 2006.

Loureiro, M.L. and W.J. Umberger. 2003. Estimating consumer willingness-to-pay for country-of-origin labeling. *J. Agric. Res. Econ.* 28 (August): 287-301.

———. 2005. Assessing preferences for country-of-origin labeled products. *J. Agric. and Appl. Econ.* 37 (1) (April): 49-63.

Lusk, J.L. 2001. Branded beef: Is it what's for dinner? *Choices.* (Spring): 27-30.

Lusk, J.L. and J. A. Fox. 2002. Consumer demand for mandatory labeling of beef from cattle administered growth hormones or fed genetically modified corn. *J. Agric. and Appl. Econ.* 34 (1) (April): 27-38.

Lusk, J.L., J.A. Fox, T.C. Schroeder, J. Mintert, and M. Koohmaraie. 2001. In-store valuation of steak tenderness. *Am. J. Agric. Econ.* 83 (August): 539-550.

Lusk, J.L., J. Roosen, and J.A. Fox. 2003. Demand for beef from cattle administered growth hormones or fed genetically modified corn: A comparison of consumers in France, Germany, the United Kingdom, and the United States. *Am. J. Agric. Econ.* 85 (1) (February): 16-29.

Marsh, T.L., T.C. Schroeder, and J. Mintert. 2004. Impacts of meat product recalls on consumer demand in the USA. *Appl. Econ.* 36: 897-909.

McGarry-Wolf, M. and A.J. Thulin. 2000. A target consumer profile and positioning for promotion of a new locally branded beef product. *J. Food Distrib. Res.* 32 (1): 193-197.

McGuirk, A., P. Driscoll, J. Alwang, and H. Huang. 1995. System misspecification testing and structural change in demand for meats. *J. Agric. Res. Econ.* 20: 1-21.

Miller, M.F., M.A. Carr, C.B. Ramsey, K.L. Crocket, and L.C. Hoover. 2001. Consumer thresholds for establishing the value of beef tenderness. *J. Anim. Sci.* 79: 3062-3068.

Mintert, J., T. Schroeder, and T. Marsh. 2002. Focus on beef demand. In: *Managing for Today's Cattle Market and Beyond.* A publication of the Western Extension Marketing Committee. March, pp. 1-4. Available online at: cals.arizona.edu/arec/wemc/TodaysCattlePub.html#2002update. Accessed January 17, 2006.

Nayga, R.M. Jr. 2000. Nutritional knowledge, gender and food label use. *J. Consumer Affairs.* 34: 97-112.

Nayga, R.M. Jr., W. Aiew, and J.P. Nichols, 2005. Information effects on consumers' willingness to purchase irradiated food products. *Rev. Agric. Econ.* 27(1): 37-48.

Nelson, P. 1970. Information and consumer behavior. *J. of Pol. Econ.* 78: 311-329.

Northern, J.R. 2000. Quality attributes and quality cues. Effective communication in the UK meat supply chain. *British Food J.* 102 (3): 230-245.

Oude Ophuis, P.A.M. and H.C.M. Van Trijp. 1995. Perceived quality: A market driven and consumer oriented approach. *Food Quality and Preference.* 6: 177-183.

Piggott, N.E. and T.L. Marsh. 2004. Does food safety information impact U.S. meat demand? *Am. J. Agric. Econ.* 86 (1) (February): 154-174.

Poulsen, C.S., H.J. Juhl, K. Kristensen, A.C. Bech, and E. Engelund. 1996. Quality guidance and quality formation. *Food Quality and Preference.* 7: 127-155.

Purcell, W.D. 1998. A primer on beef demand. Research Institute on Livestock Pricing, Virginia Tech University, Blacksburg, VA, Research Bulletin, April.

———. 2001. The future of beef business, beef demand, and opportunities for producers. In: Proceedings of the 50th Annual Florida Beef Cattle Short Course. Gainesville, Florida. May. Available online at: www.animal.ufl.edu/extension/beef/documents/Short01/Purcell.htm. Accessed on January 17, 2006.

———. 2002. Prescriptions for a healthy beef industry. In: *Managing for Today's Cattle Market and Beyond.* A publication of the Western Extension Marketing Committee. March, pp. 1-13. Available online at: cals.arizona.edu/arec/wemc/TodaysCattlePub.html#2002update. Accessed January 17, 2006.

Rimal, A. 2002. Factors affecting meat preference among American consumers. *Family Econ. Nutr. Rev.* 14: 36-43.

———. 2005. Meat labels: Consumer attitude and meat consumption pattern. *Intern. J. Consumer Studies.* 29 (1) (January): 47-54.

Roosen, J., J.L. Lusk, and J.A. Fox. 2003. Consumer demand for and attitudes toward alternative beef labeling strategies in France, Germany, and the UK. *Agribusiness: An Intern. J.* 19 (1): 77-90.

Schroeder, T.C., T.L. Marsh, and J. Mintert. 2000. Beef demand determinants. Report prepared for the National Cattlemen's Beef Association. January. Available online at: www.agmanager.info/livestock/marketing/bulletins_2/industry/demand/BeefDemandDeterminants.pdf. Accessed December 12, 2005.

Shackelford, S.D., T.L. Wheeler, M.K. Meade, J.O. Reagan, B.L. Byrnes, and M. Koohmaraie. 2001. Consumer impressions of Tender Select beef. *J. Anim. Sci.* 79: 2605-2614.

Smith, C.G., J.W. Savell, H.G. Dolezal, T.G. Field, D.R. Gill, D.B. Griffin, D.S. Hale, J.B. Morgan, S.L. Northcutt, and J.D. Tatum. 1995. The National Beef Quality Audit—1995. Colorado State University, Texas A&M University, and Oklahoma State University for the National Cattlemen's Beef Association. Centennial, CO.

Steenkamp, J.B., and H.C.M. Van Trijp. 1996. Quality guidance: A consumer-based approach to food quality improvement using partial least squares. *European Rev. of Agric. Econ.* 23: 195-215.

Sunding, D.L. 2003. The role of government in differentiated product markets: Looking to economic theory. *Am. J. Agric. Econ.* 85 (3) (August): 720-724.

Thilmany, D.D., W.J. Umberger, and A.R. Ziehl. 2006. Strategic market planning for value-added natural beef products: A cluster analysis of Colorado consumers. *J. Renewable Agric. and Food Sys.* 21 (3): 192-203.

Thilmany, D., J. Grannis, and E. Sparling. 2003. Regional demand for natural beef products in Colorado: Target consumers and willingness to pay. *J. Agribusiness* 21(2, Fall): 149-165.

Tronstad, R., R. Lobo, W. Umberger, S. Nakamoto, K.R. Curtis, L. Lev, D. Bailey, R. Ward, and C.T. Bastian. 2005. Certification and labeling considerations for agricultural producers. A Publication of the Western Extension Marketing Committee and the University of Arizona. Publication #1372. September. Available online at: cals.arizona.edu/arec/wemc/certification&labeling/certificationbook print.pdf. Accessed December 12, 2005.

Umberger, W.J. 2001. Consumer willingness-to-pay for flavor in beef steaks: An experimental economics approach. PhD dissertation, University of Nebraska-Lincoln, Lincoln.

Umberger, W.J., D.M. Feuz, C.R. Calkins, and B.M. Sitz. 2003. Country-of-origin labeling of beef products: U.S. consumers' perceptions. *J. Food Distrib. Res.* 34 (3) (November): 103-116.

Wessells, C.R., R.J. Johnston, and H. Donath. 1999. Assessing consumer preferences for eco-labeled seafood: The influence of species, certifier, and household attributes. *Am. J. Agric. Econ.* 81: 1084-1089.

Wirthgen, A. 2005. Consumer, retailer, and producer assessments of product differentiation according to regional origin and process quality. *Agribusiness.* 21 (2): 191-211.

Glossary

adipose: Of or pertaining to animal fat.

adulterant: Chemical impurities or substances that by law do not belong in a food, pesticide, or other substance.

aerobic: Requiring air, where air usually means oxygen.

affinity: A natural liking; the force that causes certain elements to combine and stay combined.

ambient: Surrounding on all sides.

anaphylaxis: A severe and rapid systemic allergic reaction to a trigger substance (called an allergen) occurring after ingestion, inhalation, skin contact, or injection of the trigger substance.

anion: A negatively charged ion.

antimicrobial: A substance that kills or slows the growth of microbes, such as bacteria (antibacterial activity), fungi (antifungal activity), viruses (antiviral activity), or parasites (antiparasitic activity).

astringency: A chemical substance that tends to shrink or constrict body tissues.

axon: A long slender projection of a nerve cell, or neuron, that conducts electrical impulses away from the neuron's cell body or soma; the primary transmission lines of the nervous system, and as bundles they help make up nerves.

biological hazard: Biological substances that pose a threat to (primarily) human health.

Bos indicus: Any of several breeds of Indian cattle; a large American heat- and tick-resistant grayish humped breed evolved in the Gulf States by interbreeding Indian cattle.

Handbook of Beef Safety and Quality

doi:10.1300/5640_09

Bos taurus: European cattle, including similar types from Africa and Asia.

branded beef program: A method of marketing beef items by emphasizing specific characteristics such as breed-type, quality, leanness, production methods, etc. Examples include: Certified Angus Beef, Sterling Silver, Laura's Lean Beef, and Certified Hereford Beef.

bung: The portion of an animal that includes the perineal skin, anus, rectum, and external sphincter muscle, which is removed and tied to prevent fecal leakage during evisceration.

carriage: In the case of contamination, fecal matter on the legs of cattle, carried from one place to another.

colonized: An established group of similar bacteria growing in or on a culture medium.

contamination: Something that causes a product to be infected, corrupted, or polluted after contact with it.

cross-contamination: When bacteria is spread between food, surfaces, and/or equipment.

cutaneous trunci: A relatively thin muscle near the skin in the live animal beginning on the shoulder and ending in the flank.

desiccated: Completely dried.

dissociate: To undergo a reversible or temporary breakdown of a molecule into simpler molecules or atoms; to disperse or spread out.

e-beam irradiation: A stream of high-energy electrons used in food irradiation, sterilization, welding, imaging, etc.

electronic identification: Use of bar codes and/or radio frequency transponders for identifying livestock.

electrostatic: Produced by or relating to static electricity.

emulsification: To convert two or more liquids into an emulsion, or become an emulsion.

Enterobacteriaceae: A large family of bacteria, including pathogens such as *Salmonella* and *Escherichia coli*.

enterotoxin: A toxin produced by bacteria that is specific for intestinal cells and causes the vomiting and diarrhea associated with food poisoning.

evisceration: The act of removing the bowels or viscera.

exsanguination: Fatal process of total blood loss used as a method of harvest in the meat industry.

extrapolation: Using historical data to try to determine what will happen in the future.

fabricated: Manufactured; in the case of beef carcasses, manufacturing carcasses into a more easy-to-manage and sellable product.

facultative: Capable of functioning under varying environmental conditions.

foreign matter: Unwanted or undesirable material present in foods or chemicals; may include packing materials inadvertently (or deliberately) included in the product, plant or meat products that should have been removed in manufacture or processing, vermin remains, stones, grit, sand, etc.

frock: A long loose outer garment, usually white, that is worn over a worker's clothing.

FSIS: Food Safety Inspection Service; the government agency within the United States Department of Agriculture that is responsible for federal meat inspection.

GRAS: Generally Regarded As Safe.

HACCP: Hazard Analysis Critical Control Point; a production control system for the food industry designed to prevent rather than catch potential hazards; identifies where potential contamination can occur (the critical control points or CCPs) and strictly manages and monitors these points as a way of ensuring that the process is in control and that the safest product possible is being produced.

hardboned carcasses: A carcass exhibiting one or both of the following characteristics: 30 percent ossification of the first three full thoracic vertebrae and/or a chronological age of 42 months or older.

high-concentrated diets: A feed that is made up of over 80 percent non-roughage.

hydrophilic: Having an affinity for water; readily absorbing or dissolving in water.

hydrophobic: Repelling, tending not to combine with, or incapable of dissolving in water.

hydrostatic pressure: The pressure due to the weight of a fluid.

inferential limitations: Limits that are inferred by the use of other substantiated data.

innervated: To supply an organ or a body part with nerves.

inoculum: A medium containing organisms, usually bacteria or a virus, that is introduced into cultures or living organisms.

integument: An outer protective covering such as the skin of an animal or a cuticle or seed coat or rind or shell.

intervention: An interference.

intoxication: To become poisoned.

isoelectric point: The pH at which a molecule carries no net electrical charge.

kinetic energy: Energy that a body has as a result of its speed or energy of motion.

malt: Mucosa-associated lymphoid tissue; scattered along mucosal linings, measuring roughly 400 m^2 to protect the body from an enormous quantity and variety of antigens (e.g., tonsils, Peyers patches within the small intestine).

micronutrient: Essential elements only needed by life in small quantities.

multiple hurdles: Using more than one intervention or technology to interfere with microbial growth.

neoplasia: Abnormal, disorganized growth in a tissue or organ, usually forming a distinct mass.

neurodegenerative: A disorder caused by the deterioration of certain nerve cells (neurons).

norovirus: Small viruses that can cause gastroenteritis, or "stomach flu" (the illness is not related to the flu, which is a respiratory illness caused by the influenza virus); being viruses and not bacteria or parasites, the symptoms of infection are not relieved by the use of antibiotics.

nutrient value: The amount of a chemical compound (such as protein, fat, carbohydrate, vitamins, or minerals) that make up foods.

ozone: A molecular form of oxygen, the molecule consisting of three oxygen atoms instead of the more stable diatomic O_2; can be used for bleaching substances and for killing bacteria.

packer grid: A method of assigning value to beef carcasses based on yield and/or quality grade with discounts for carcasses exhibiting characteristics such as blood splash, hardbone, dairy type, bruising; and premiums paid for carcasses exhibiting superior carcass traits.

pasteurization: The process of heating food for the purpose of killing harmful organisms such as bacteria, viruses, protozoa, molds, and yeasts.

pathogens: Biological agent that causes disease or illness to its host.

Peyers patches: Lymphoid nodules in the walls of the small intestines.

phagocytes: Cells that serve to remove foreign bodies and thus fight infection; in vertebrates these include larger macrophages and smaller granulocytes, types of blood cells.

pluck: Trachea, lungs, and heart.

primals: The four major cuts of the beef carcass including the round, rib, loin, and chuck.

probiotics: Live microorganisms which when administered in adequate amounts confer a health benefit on the host.

purveying: To furnish or provide, as with a convenience, provisions, or the like.

quality grade: A composite evaluation of degree of marbling and degree of maturity that affect palatability of meat (tenderness, juiciness, and flavor).

quaternary: Consisting of or especially arranged in sets of four.

rectoanal junction: The junction of the rectum with the anal canal.

resistant determinants: Having exact limits that give the ability to ward off bacteria.

rework: A meat item that has not met certain specifications and is, therefore, sent back through processing in an attempt to correct any defects that might be present.

RTE: Ready-to-eat; food products that are prepared in advance and can be eaten as sold.

septicemia: A systemic disease caused by pathogenic organisms or their toxins in the bloodstream; also called *blood poisoning*.

shelf life: The length of time a product may be stored without becoming unsuitable for use or consumption.

starter cultures: Microorganisms that are intentionally added to produce a desired outcome in the final product, most often through their growth and "fermentation" of the food product, i.e., cheese, sausages.

subacute: Somewhat or moderately acute; between acute and chronic.

suicide substrate: A substrate that when acted upon by an appropriate enzyme is converted into a product that irreversibly inactivates the enzyme usually by covalent modification.

total plate counts: Determination of the total number of microorganisms found within a demarcated region of a slide known to hold a certain volume of culture.

triclosan: A potent wide-spectrum antibacterial and antifungal agent; (5-chloro-2-(2,4-dichlorophenoxy)phenol).

umami: Savoriness; the taste sensation produced by the free glutamates commonly found in fermented and aged foods, for example parmesan and roquefort cheeses, soy sauce, and fish sauce; the glutamate taste sensation is most intense in combination with sodium.

value-added: Additional value created at a particular stage of production such as marinating a less-tender or undervalued cut.

violative residues: Residues resulting from the use of animal drugs and pesticides or from incidents involving environmental contaminants that are outside the tolerances established by the Food and Drug Administration, Environmental Protection Agency, and FSIS.

viscera: Internal organs, collectively known as offal.

yield grades: Estimation of the amount of boneless, closely trimmed retail cuts from the high-value parts of the carcass—the round, loin, rib, and chuck.

zero tolerance: Generally refers to the government standard for no visible contamination on carcasses; no degree of hazard is acceptable.

Index

Page numbers followed by the letter "f" indicate figures; those followed by the letter "t" indicate tables.

Abraham, H. C., 132, 138, 139
Acebron, L. B., 205
Acetic acid, 56-57
Acidified sodium chlorite (ASC), 58
Acuff, G. R., 54, 57, 58, 59
Age
  high-concentrate diets and, 164
  impact on palatability, 167
  postharvest aging, 171-172
Agricultural Marketing Service (AMS), 98, 127, 200
Allergens
  big 8, 45
  as chemical hazard, 44
  control of, in plants, 45
  defined, 44
  immune responses, 44-45
Alwang, J., 194
AMDUCA (Animal Medicinal Drug Use Clarification Act), 104-107
American Association of Bovine Practitioners, 119
American Meat Science Association (AMSA)
  flavor measurement, 159
  juiciness scale, 155
  Sensory and Cookery Guidelines for Meat, 148
  sensory evaluation of quality, 148, 151-152
  tenderness, 157
Andrews, A. H., 143
Angus beef, 161, 163
Animal Identification System (NAIS), 122
Animal Medicinal Drug Use Clarification Act (AMDUCA), 104-107
Antibiotics in beef production, 43-44
Antimicrobial drug resistance
  complexity of issues, 24-25
  in-feed drugs, 24
  non-O157 STEC, 23-24
  Salmonella, 23, 24
Antimicrobial interventions
  carcass treatment, 42
  cetylpyridinium chloride, 60-61
  chemical dehairing, 62-64
  chlorinated compounds, 58-59
  combinations, 64-65
  future/potential interventions, 66-67
  knife trimming, 42, 64
  lactoferrin, 61
  organic acids, 56-57
  ozonated water, 59-60
  peroxyacetic acid, 57-58
  physical attributes, 65-66
  rinsing with antimicrobial compounds, 55, 65
  sodium diacetate, 69, 175
  steam pasteurization, 20, 62
  steam vacuuming, 61-62
  water washing, 54-55. *See also* Water washing
APC, 63, 65
Aromatics, 159, 174
Arthur, T. M., 60-61
ASC (acidified sodium chlorite), 58
Ascophyllum nodosum, 26

Handbook of Beef Safety and Quality

doi:10.1300/5640_10

Attributes of beef
credence attributes, 198, 201, 206, 207-208
cues. *See* Cues
defined, 198
experience attributes, 198, 206
flavor. *See* Flavor attribute
free range, 197, 206
grass-fed, 197
impact on demand, 3-4
juiciness, 155-157, 164-165, 174
lean, 197
natural, 197, 200-201, 206
organic, 197, 200-201, 206
process attributes, 198-199
rankings and ratings, 203-204
search attributes, 198
sensory. *See* Sensory attributes
texture, 132-133, 179
AUS-MEAT, 137
Austin, Nancy, 85
Australia
dentition in classification system, 143
grading system, 136-137
Avian Influenza (Bird Flu), 91

Bailey, D., 201
Baker, J. J., 166
Ballots
consumer sensory evaluation ballot, 156
Meat Descriptive Sensory Evaluation, 149
Spectrum Method, 153
Bastian, C. T., 201
BCS (body condition score), 114
Bechtel, P. J., 177
Becker, T., 197, 198
Beef Export Verification (BEV) Program, 96
Beef industry
changing industry, 16
consumer driven industry, 4
consumer product business, 16
end-user responsive industry, 4
foodborne pathogen concerns, 11-15, 12t
production driven industry, 4
relationship business, 16-17
Beef Industry Food Safety Council (BIFSCo), 12-13
Beef Quality Assurance (BQA) Program
antimicrobial intervention, 44
foodborne pathogens, 122
future issues, 122-123
goals, 103
history, 101-103
total quality management, 91
Beekman, D. D., 170
Bégin, A., 56, 57
Belk, K. E., 60, 174
Belk, Keith, 32
Berry, B. W., 139
BEV (Beef Export Verification) Program, 96
BIFSCo (Beef Industry Food Safety Council), 12-13
Big 8 allergens, 45
Biological hazards
dangers of, 43
diseases. *See* Diseases
post-fabrication processing, 67-68
statistics of incidence, 47
Bird Flu (Avian Influenza), 91
Blade tenderization, 172
Blood splash, 168
Bobrow, M., 52-53
Body condition score (BCS), 114
Bos indicus breeds, 160-161, 162t, 163, 166
Bos Taurus breeds, 160-161, 162t, 163
Bosilevac, J. M., 60-61, 63
Botulinum toxin, 174
Bovine Respiratory Disease (BRD), 116, 120-122
Bovine spongiform encephalopathy (BSE, mad cow disease)
disease prevention, 12-13, 53-54
history of, 52
international trade and, 91
Japanese imports of beef, 96
maturity of cattle and, 140
media coverage, 193
prevention during slaughter, 41
prion, 52-53
sources of, 47
symptoms, 52
testing for disease, 53

Boyle, A. E., 64
BQA. *See* Beef Quality Assurance (BQA) Program
Branson, R. E., 138
Brashears, M. M., 28
Brashears, Mindy M., 19
Bray, R. W., 126
BRD (Bovine Respiratory Disease), 116, 120-122
Briskey, E. J., 126
Brown, E. G., 166
Bruce, M., 52-53
Bryan, T. M., 166
Bryant, J., 62
BSE. *See* Bovine spongiform encephalopathy (BSE, mad cow disease)
Buege, D. R., 54, 59
Bulk Density Theory, 130
Business. *See* Beef industry
Buttons, 133
Buyck, M. J., 54, 59, 140, 158, 160
Byrnes, B. L., 206

Calcium chloride, 175-176
Calicivirus (Norwalk virus), 6, 52
Calpains, 171, 175
Calpastatin, 160-161, 163, 171-172
Calving, 115, 116
Campylobacter jejuni
- as cause of diarrhea, 7-8
- control of, 8
- generally, 6
- incidence and death rate, 9t, 10
- incidence decrease, 10
- prevalence on carcasses, 10t
- sources of, 8, 47, 49
- symptoms, 49

Canadian yield grades, 134-135, 135t
Carballo, J., 69
Carcasses
- certification programs, 98, 200-201, 207
- *E. coli* O157:H7 contamination, 20-21, 26-27, 27f
- quality grades. *See* Grading of carcasses
- slaughter procedures, 41-42

Care and handling of beef cattle
- disease prevention, 115-116
- downer cattle, 118-119
- emergency procedures, 120
- euthanasia, 118-119
- feeding and nutrition. *See* Feeding cattle
- feedlot heat-stress, 119-120
- handling cattle, 116-117
- health care, 115-116
- marketing cattle, 118
- overview, 113

Carpenter, Z. L., 132, 138, 139
Carr, B. T., 155, 158-159
Carr, M. A., 206
Carr, T. R., 177
Carrascosa, A. V., 69
Carstens, G. E., 165, 166
Casas, E., 163
CAST (Council of Agriculture and Science and Technology), 15
Castillo, A., 58
Castration, 115-116
Caul, J. F., 158-159
Centers for Disease Control and Prevention (CDC), 6
Central Location Consumer Tests, 154
Certification for beef, 98, 200-201, 207
Cetylpyridinium chloride (CPC), 60-61
Chemical dehairing, 62-64
Chemical hazards
- as cause of illness and injury, 43
- feedlot assessment, 101-102
- hormones and antibiotics, 43-44
- nonmeat ingredients, 44, 68
- in plant, 44
- post-fabrication processing, 67-68
- sources, 43

Chemical treatments, 42
Chicago beef study, 205
Chicken, *Campylobacter jejuni* contamination, 8. *See also* Poultry
Chilling. *See* Cold-shortening
Chlorinated compounds, 58-59
Chlorine dioxide, 59
Cholesterol as health concern, 194
Chutes for restraint, 112-113, 117
Citric acid, 56-57, 69
Civille, Gail Vance, 151, 155, 158-159, 172

Clayton, R. P., 63
Clostridium perfrigens, 10t
Codes Guidelines, 93
Codex Alimentarius, 92-93, 93t
Codex Codes of Practice, 93
Cold tolerance, 166
Cold-shortening, 164, 169-170, 171
Colmenero, F. J., 69
Color in grading, 131-133, 137
Commodity marketing, 97
Competitive exclusion, 27-28
Connective tissue amount, 157-158
Consumer
  carcass quality and acceptability, 137-140
  consumer product business, 16
  consumer-driven industry, 4
  cues, role in perceptions, 199-203
  demands, 188-189
  demographics, 194-196, 195f, 207-208
  differences from customer, 4
  food demand pyramid, 195f
  market share decreases, 187-188
  origin, traceability, and production information, 206-208
  prices and expenditures, 191-192
  sensory evaluation, 154
  sensory evaluation ballot, 156
Cooling during slaughter, 41
Corn in finishing diets, 165
Costs
  chemical dehairing, 64
  of foodborne pathogen disease, 47
  impact on quality, 89, 90
  of interventions during slaughter, 42
  price of product, 4, 191-192
  of reducing *Escherichia coli* O157:H7, 11, 12t
  steam vacuum sanitizer, 62
  steam-pasteurization systems, 62
Council of Agriculture and Science and Technology (CAST), 15
Country-of-Origin Labeling (COOL), 207
Cowman, G. L., 109
CPC (cetylpyridinium chloride), 60-61
Crawford, A. M., 163
Credence attributes, 198, 201, 206, 207-208
Credibility of extrinsic cues, 200
Crocket, K. L., 206
Cross, H. R., 126, 130, 132, 138, 139
Cross-contamination
  of allergens, 45
  as cause of *Campylobacter* spp., 49
  as cause of *Listeria monocytogenes*, 47
Cryptosporidium, 9t
Cues
  extrinsic cues, 199-203, 202f
  intrinsic cues, 199-203, 202f, 204-205
  marbling, 204-205
  rankings and ratings, 204
  role in consumer perceptions, 199-203
  Total Food Quality Model, 201-202, 202f
Cullen, N. G., 163
Curley, K. O., 166
Curtis, K. R., 201
Customer, differences from consumer, 4
Cutter, Catherine N., 39, 65, 66, 71

Davis, G. W., 132, 138, 139
Dehorning, 115-116
Delaney Amendment, 104
Demand
  consumer demands, 188-189
  consumer food demand pyramid, 195f
  defined, 188
  demand drivers, 3-4
  demographics, 194-196, 195f, 207-208
  factors in, 3
  fat and cholesterol level concerns, 194
  increases and decreases, 189
  index, 189-190, 190f
  media coverage of food safety events, 193-194
  models, 190-196, 195f
  overview, 187-188, 208
  quantity demanded, 189
  relative meat prices and consumer expenditures, 191-192
  safety and health concerns, 192-194

Deming, W. Edwards, 89, 90-91, 95
Demographics of consumer demand, 194-196, 195f, 207-208
Dentition
for aging cattle, 140
deciduous teeth, 140-141
eruption of permanent teeth, 141, 142f, 142t
milk teeth, 140-141
permanent teeth, 141, 141f
skeletal maturity and, 141-143
Detection equipment, 46-47
DeVol, D. O. L., 177
Dexter, D. R., 109
Diarrhea caused by *Campylobacter jejuni*, 7-8
Dickson, J. S., 54, 55, 59
Diets, high-concentrate, 164-165
Dikeman, M. E., 143
Direct-fed microbials
competitive exclusion, 27-28
Enterococcus, 29-30, 33
Lactobacillus acidophilus, 28-29
probiotics, 27
Diseases
Avian Influenza (Bird Flu), 91
Bovine Respiratory Disease, 116, 120-122
BSE. *See* Bovine spongiform encephalopathy
calicivirus (Norwalk virus), 6, 52
Campylobacter jejuni. See Campylobacter jejuni
Clostridium perfrigens, 10t
Cryptosporidium, 9t
Enterobacteriaceae, 63
Escherichia coli, non-0157 STEC, 23-24
Escherichia coli O157:H7. *See Escherichia coli* O157:H7
Foot and Mouth Disease, 91
Guillian-Barré syndrome, 49
hepatitis A, 47, 52
Listeria monocytogenes. See Listeria monocytogenes
non-O157 STEC, 23-24
Noroviruses, 47, 52
Norwalk virus, 6, 52
outbreaks, 193
prevention of, 115-116
Pseudomonas fluorescens, 57
Diseases *(continued)*
Reiter's syndrome, 49
Salmonella. See Salmonella
Shigella. See Shigella
Staphlococcus aureus. See Staphlococcus aureus
statistics of incidence, 6, 47
transmissible spongiform encephalopathy, 52
Vibrio, 9t
viruses. *See* Viruses
Yersinia enterocolitica. See Yersinia enterocolitica
Documentation
drug administration, 108-109
injections, 113
quality/management practices, 6
Domestic markets, quality demands, 94-95
Dopico, D. C., 205
Dorsa, W. J., 66
Downer cattle, 118-119
Driscoll, P., 194
Driving aids, 117
Drugs
antibiotics in beef production, 43-44
extra-label usage, 104-107
immunomodulation, 30-33
inspection for residue, 43-44
neomycin sulfate, 31-33
over-the-counter, 105-106
prescriptions, 106
regulations, 103-105
residue avoidance, 107-109
resistance to. *See* Antimicrobial drug resistance
sodium chlorate, 31
withdrawal time, 108-109

Ecchymosis, 168
Elam, N. A., 28
Elder, R. O., 20
Elder, Robert, 32
Electrical stimulation, 168-170
Emergency procedures, 120
Employee training
about quality products, 91
allergen control, 45
competent instruction, 90

Employee training *(continued)*
employees as source of viral contamination, 52
foreign objects and personal equipment, 46-47
physical safety training, 46
prevention of *Listeria monocytogenes,* 50
in sanitation procedures, 44
Shigella prevention, 51-52
Endomysium, 167
End-user responsive industry, 4
Enterobacteriaceae, 63
Enterococcus, 29-30, 33
Environmental impact on cattle, 165-166
Equipment
antimicrobial enhancement, 71
machinery hazards, 46
object detection equipment, 46-47
Escherichia coli, non-O157 STEC, 23-24
Escherichia coli O157:H7
antimicrobial intervention, 57-58, 60, 65, 66
from carcasses and hides, 20-21, 26-27, 27f
cattle as carrier, 5
chemical dehairing, 63
contamination during hide removal, 41
control of, 7, 25-26, 33
costs of reducing prevalence of, 11, 12t
damage caused by, 7
direct-fed microbials, 27-30
factors of contamination, 13
generally, 6
ground beef contamination, 7, 11, 12t, 22f
HACCP implementation, 40
human incidence, 22, 23f
incidence and death rate, 9t, 10
incidence by year, 10, 11t
incidence decrease, 10-11
interventions, 25-26
irradiation, 66
knife trimming intervention, 64
media coverage, 193
neomycin sulfate control, 31-33
organic acid effectiveness, 57
Escherichia coli O157:H7 *(continued)*
preharvest control, 20-21, 25
prevalence on carcasses, 10t
processing interventions, 69
sodium chlorate control, 31
sodium metasilicate intervention, 66
sources of, 7, 47, 48
steam pasteurization, 20, 62
susceptible individuals, 48
symptoms, 47
vaccine, 30-31
Espitia, F. Danielle, 125
EUROP, 136
European Union grading system, 136
Euthanasia, 118-119
Evisceration, 41
Experience attributes, 198, 206
Exsanguination, 40-41, 168
Extra-label drug use, 104-107
Extrinsic cues, 199-203, 202f

Falkenberg, S. M., 166
FARAD (Food Animal Residue Avoidance Databank), 108-109
Fat
factor in consumer demand, 194
impact on flavor and juiciness, 164-165
Fatty acids in muscle, 177, 178t
FDA (Food and Drug Administration), 104, 200
Feces
E. coli O157:H7 contamination, 21, 29-30
neomycin sulfate control of *E. coli,* 32
Federal Food and Drug Act, 103
Federal Food, Drug, and Cosmetic Act (FFD&C), 103-104
Feed additives, extra-label use, 105
Feeding cattle
high-concentrate diets, 164-165
in-feed drugs, 24, 32
nutrition requirements, 113-114
Feedlot
chemical hazard assessment, 101-102
diets, 114
E. coli O157:H7 prevalence, 25-26

Feedlot *(continued)*
heat-stress, 119-120
impact of BRD, 121
study of *Escherichia coli* O157:H7, 20-21
Feldkamp, T. J., 206
Fernandez, P., 69
FFD&C (Federal Food, Drug, and Cosmetic Act), 103-104
Field, Thomas G., 85
Firmness, in grading, 132-133
Flavor attribute
aging impact, 172
Flavor and Texture Descriptive Attribute Evaluation, 150-154, 179
generally, 158-160
high-concentrate diets impact, 164-165
muscle flavor desirability ratings, 177, 177t, 179
sodium lactate ingredient, 174-175
sodium phosphates, 173-174
warmed-over flavors, 150t
FMD (Foot and Mouth Disease), 91
Food Animal Residue Avoidance Databank (FARAD), 108-109
Food and Drug Administration (FDA), 104, 200
Food handlers. *See* Employee training
Food Marketing Institute (FMI), "Trends in the United States", 5
Food Safety and Inspection Service (FSIS)
credibility for consumers, 200
dentition, 140
drug residue inspection, 43-44
E. coli O157:H7 in ground beef, 69
food safety impacts, 193-194
history, 101
removal of contamination, 42
Foodborne pathogens
beef industry concerns, 11-15, 12t
diseases. *See* Diseases
Hazard Analysis and Critical Control Point, 13-14
preharvest BQA issues, 122
prevention, 13-14
FoodNet, 9
Foot and Mouth Disease (FMD), 91
Forages, 114
Fox, D. G., 166
Fox, J. A., 206, 207
Fox, J. T., 166
Free range attribute, 197, 206
FSIS. *See* Food Safety and Inspection Service (FSIS)
Fung, D. Y. C., 64

Galyean, M. L., 28
Generally recognized as safe (GRAS) compounds, 55, 61, 67
Genetics
GeneSTAR test, 163
for tenderness, 163-164
Genho, P. C., 166
George, M. H., 109
Gerthoff, T., 170
Gill, C. O., 62
Glock, R. D., 109
Glycolysis, 170, 171, 176
Glycolytic muscle fibers, 176
GMPs (good manufacturing practices), 40
Godert, M., 52-53
Good manufacturing practices (GMPs), 40
Goodson, K. J., 58
Gorman, B. M., 63, 65
Grading of carcasses
background, 126-128
BRD treatment impact, 121
carcass quality and consumer acceptability, 137-140
classification system, 126
color, 131-133, 137
current standards, 128
cutability standards (Yield Grades), 127-128
dentition, 140-143, 141f, 142f, 142t
firmness, 132-133
function of quality grades, 125-126
hierarchy, 133
international standards, 91-93, 93t, 134-137, 135t
marbling. *See* Marbling
maturity, 131, 132-134, 134t, 135f
maturity groups, 138
muscle specific, 130-131
overview, 125, 143-144

Grading of carcasses *(continued)*
during processing, 42
quality grades, 139-140
skeletal maturity, 133, 140, 141-143
texture and firmness, 132-133
U.S. standards, 133-134, 134t
Window of Acceptability, 130, 131f
Grain-fed cattle, 165
GRAS (generally recognized as safe), 55, 61, 67
Grass-fed attribute, 197
Griffin, D. B., 138
Griffin, D. D., 13
Ground beef
BSE prevention, 54
E. coli O157:H7 contamination, 7, 11, 12t, 22f
prevalence of *Escheria coli* O157:H7, 11, 12t
processing interventions, 69
Growth hormones, 43
Grunert, K. G., 201
Guidelines for Care and Handling of Beef Cattle, 113
Guillian-Barré syndrome, 49
Gwartney, B. L., 63

Hair
chemical dehairing, 62-64
as physical hazard, 46
Handling cattle, 116-117
Hardin, M. D., 57
Hardin, Margaret D., 39
Harris, J. J., 126
Hazard Analysis and Critical Control Point (HACCP)
cleaning carcasses, 42
Food Safety and Inspection Service, 42. *See also* Food Safety and Inspection Service (FSIS)
foodborne pathogens, 13-14
implementation of procedures, 15, 16t, 40
principles, 14
TQM implementation, 91
Hazards. *See* Safety hazards
Health concerns and demand for beef, 192-194. *See also* Diseases
Heaton, M. P., 163
Heat-stress, 119-120, 166
Heinrich, P. E., 109
Hemoglobin, 132
Henning, William R., 39
Hepatitis A, 47, 52
Herbs and spices, 46, 68
Hereford beef, 161, 163
Hides of cattle
chemical dehairing, 62-64
CPC decontaminant, 60-61
E. coli O157:H7 contamination, 7, 20-21, 26-27, 27f, 29-30
hair as physical hazard, 46
neomycin sulfate control of *E. coli,* 32
removal during slaughter, 41
High-concentrate diets prior to slaughter, 164-165
Hoenecke, M. E., 170
Hoffmann, R., 202-203
Holley, R. A., 56, 57
Holloway, J. W., 166
Hoover, L. C., 206
Hormones, 43
Horn, G. H., 52-53
Hsia, C. J., 194
Huang, H., 194
Huerta-Montauti, Diana, 125
Hump height, 137
Hyndman, D. L., 163

IAMP (Institute of American Meat Packers), 127
Identification of cattle, 108, 122
Illness. *See* Diseases
Immobilization during harvest, 168
Immunomodulation, 30-33
Income, impact on demand, 195-196, 201
Industry. *See* Beef industry
In-feed administration of drugs, 24, 32
In-Home Consumer Tests, 154
Injection needles, 46
Injection solutions
calcium chloride, 175-176
meat enhancement, 173-176
potassium lactate, 69, 174
sodium chloride, 173-174
sodium diacetate, 69, 175
sodium lactate, 69, 174-175
sodium phosphates, 173-174

Injection-site lesions
animal restraint, 112-113
cause of, 109
identification, 102
intramuscular administration, 110-111, 111f
needles for injections, 111
persistence of, 109-110, 110f
record-keeping, 113
subcutaneous administration, 110-111, 112f
volume of drugs, 112
Inspections, 41, 43-44
Institute of American Meat Packers (IAMP), 127
Insurance Theory, 130
International Beef Quality Audit, 95-97
International grading standards, 134-137, 135t
International markets
Beef Export Verification (BEV) Program, 96
International Beef Quality Audit, 95-97
Quality System Assessment, 96-97
standards, 91-93, 93t
strengths of U.S. beef, 95
weaknesses of U.S. beef, 95-96
International Organization of Standardization
ISO 9000, 92, 98
quality, defined, 197
Interventions
antimicrobial. *See* Antimicrobial interventions
irradiation, 66
sodium metasilicate. 67
types, 42
ultraviolet light, 66-67, 71
washing the carcass, 41, 42
Intramuscular administration of drugs, 110-111, 111f
Intrinsic cues, 199-203, 202f, 204-205
Irradiation, 66
ISO 9000, 92, 98

Jackson, J. D., 194
Jantschke, M., 45-46
Japan
grading system, 136
imports of beef, 96
Jaroni, D., 28
Johnsen, P. B., 151, 172
Jones, B. K., 170
Juiciness attribute
high-concentrate diet impact, 164-165
overview, 155-157
sodium lactate ingredient, 174

Kang, D., 69
Kansas City beef study, 138
Kastner, C. L., 54, 59, 64
Katsuyama, A. M., 45-46
Kay, Steve, 11, 12t
Keele, 163
Kefauver-Harris Amendment, 104
Keisler, D. H., 166
Killinger-Mann, K., 28
Kinnucan, H. W., 194
Kirschten, D. P., 166
Knife trimming, 42, 64
Koh, Y. O., 170
Koohmaraie, M.
antimicrobial intervention. 60, 63, 69
dentition and skeletal maturity, 142
gene markers, 163
postharvest sensory attributes. 170, 176
tenderness of beef, 206

Labeling of beef, 207
Lactic acid, 56-57, 69
Lactobacillus acidophilus, 28-29
Lactoferrin, 61
Lamb, calcium chloride infusion, 175
Lawrence, T. E., 143
Laws. *See* Legislation
Lean
attribute, 197
color, 137
maturity, 133
Legislation
Animal Medicinal Drug Use Clarification Act, 104-107
Delaney Amendment, 104

Legislation *(continued)*
  drugs, 103-105
  Federal Food and Drug Act, 103
  Federal Food, Drug, and Cosmetic Act, 103-104
  Kefauver-Harris Amendment, 104
  purpose of, 105
  United States Agricultural Products Inspection and Grading Act, 126
Lesions
  from injections. *See* Injection-site lesions
  pulmonary, 121
Lev, L., 201
Lipids
  in muscle, 177, 178t
  oxidation, 172, 174-175
Listeria monocytogenes
  affected people, 8-9
  control of, 9
  disease prevention, 50
  frequency of infection, 9
  incidence and death rate, 9t
  prevalence on carcasses, 10t
  in ready-to-eat products, 50, 70-71
  sources of, 9, 47, 50
  steam pasteurization, 62
  symptoms, 50
Littlefield, V. G., 63
Lobo, R., 201
Loneragan, G. H., 28
Loneragan, Guy H., 19
Lorenzen, C. L., 140, 158, 160
Loureiro, M. L., 204, 207
Lubrication Theory, 130
Lucia, L. M., 57, 58
Lunt, D. K., 165
Lusk, J. L., 204, 206, 207
Lyons, B. G., 159

M. iliocostalis, 137
M. longissimus thoracis
  Japanese meat grading score, 136
  marbling score, 129, 133, 134-135
  pH and lean color, 137
  Warner-Bratzler shear force values, 162t
Machinery. *See* Equipment
Mad cow disease. *See* Bovine spongiform encephalopathy (BSE, mad cow disease)
MALT (mucosal associated lymphoid tissue), 53
Mann, J. E., 28
Marbling
  carcass quality and consumer acceptability, 138-139
  consumer preferences, 204-205
  defined, 129, 133
  high-concentrate diets, 164
  maturity and quality, 135f
  as negative quality cue, 200
  palatability ratings, 129-130, 129f
  quality cues, 204-205
  quality grades, 135
  theories, 130
Marcus, Stanley, 85
Marketing
  Agricultural Marketing Service, 98, 127, 200
  commodity marketing, 97
  domestic markets, 94-95
  Food Marketing Institute, 5
  handling of cattle, 118
  international. *See* International markets
  market classification system, 126
  market share decreases, 187-188
  value-added marketing, 97-98
Marsden, J. L., 54, 59
Marsh, B. B., 170, 171
Marsh, T. L., 191, 192-193, 194, 195
Mason, Carrie L. Adams, 125
Maturity
  in grading, 131, 132-134, 134t, 135f
  maturity groups and consumer acceptability, 138
McAdams, Jim, 11
McGuirk, A., 194
McKeith, F. K., 177
McLean, A., 52-53
McMullen, L. M., 10
Meade, M. K., 206
Meat Descriptive Attribute Sensory Evaluation, 148-150, 179
Meat Descriptive Sensory Evaluation, 148-150
Meat enhancement, 173

Media reports on food safety events, 193-194
Meilgaard, M., 155, 158-159
Melton, S. L., 165
Microbials, 29. *See also* Direct-fed microbials
Miles, R. S., 177
Miller, M. F., 206
Miller, R. K., 140, 158, 160, 165, 166
Miller, Rhonda, 147
Mintert, J., 191, 192, 194, 195, 206
Models
  demand, 190-196, 195f
  Hoffmann's food quality model, 202-203
  Total Food Quality Model, 201-202, 202f
Moist heat treatments, 42
Montgomery, T. H., 143
Moore, S. A., 166
Morgan, J. B., 54, 59, 63, 65, 109
Morris, C. A., 163
Mossel, D. A. A., 56
M.rhomboideus, 137
Mucosal associated lymphoid tissue (MALT), 53
Mumford, Herbert, 126
Murphey, C. E., 132, 138, 139
Muscle
  flavor desirability ratings, 177, 177t, 179
  lipid and fatty acid composition, 178t
  Muscle Profiling, 177
  muscle specific, 130-131
  palatability, 176-179, 177t, 178t
  tenderness, 157, 158
Myoglobin, 132

Nakamoto, S., 201
National Animal Identification System (NAIS), 122
National Beef Quality Audit, 94
National Cattlemen's Beef Association (NCBA)
  BQA Program. *See* Beef Quality Assurance (BQA) Program
  Guidelines for Care and Handling of Beef Cattle, 113
National Consumer Retail Beef Study, 128, 138-139
National Livestock Marketing News Service, 126
National Research Council (NRC), *Nutrient Requirements of Beef Cattle,* 113-114
Natural beef, 197, 200-201, 206
NCBA. *See* National Cattlemen's Beef Association (NCBA)
Needles for injections, 111
Neely, T. R., 140, 158, 160
Neomycin sulfate, 31-33
Nickelson, R. II, 54, 59
Nitrates, 44
Nitrites, 44, 68
Non-ambulatory cattle, 118-119
Non-intact beef products, 70
Nonmeat ingredients
  as chemical hazard, 44, 68
  injection solution, 173
  microbiological quality, 68
  sodium or potassium lactate, 174
Non-O158 STEC drug resistance, 23-24
Noroviruses, 47, 52
Northern, J. R., 199-200
Norwalk virus (calicivirus), 6, 52
Nou, X., 63
NP35, 28
NP45, 29
NP51, 28, 29-30, 33
NPC747, 28
NPC750, 28
NRC. *See* National Research Council (NRC)
Nutrient Requirements of Beef Cattle, 113-114
Nutrition for cattle, 113-114

Odde, K. G., 109
Offer, G., 173
Office of Markets and Rural Organization, 126
Oman, J. S., 57
Oregon Country Beef, 85
Organic acids, 56-57
Organic beef, 197, 200-201, 206
Origin information for consumers, 206-208

Ouattara, B., 56, 57
Outbreaks, food safety, 193
Overall beef tenderness, 158
Over-the-counter (OTC) drugs, 105-106
Ozonated water, 59-60
Ozone, 71

Packaging materials
  aerobically stored beef, 172
  as chemical hazard, 44
  as physical hazard, 46
  vacuum-packaging, 172
Page, B. T., 163
Pagel, L. A., 170, 171
Palatability traits, 161t, 164
Parrish, F. C. Jr., 139
Paschal, J. C., 166
Pasteurization, steam, 20, 62
Pasture-fed cattle, 165
Paterson, J., 4
Paterson, John, 3
Pathogens
  foodborne. *See* Foodborne pathogens
  history of interest in, 22-26, 22f, 23f, 27f
  preharvest control. *See* Preharvest control of pathogens
Pens of cattle. *See* Feedlot
Perimysium, 167
Perino, L. J., 143
Peroxyacetic acid, 57-58
Peroxyacid compounds, 71
Peters, T., 88, 89
Peters, Tom, 85, 87
pH
  measurement, 137
  postharvest impact on tenderness, 170-171
  sodium diacetate ingredient, 175
  water-holding capacity, 173-174
Phebus, R. K., 64
Philadelphia beef study, 138, 139
Physical hazards
  bone, hair, insects, 46
  as cause of illness and injury, 43
  as cause of personal injury, 45
  caused by employees, 46
Physical hazards *(continued)*
  detection equipment, 46
  post-fabrication processing, 67-68
  prevention of, 46-47
  during slaughter, 46
Piette, G.J.-P., 56, 57
Piggott, N. E., 192-193, 194
Pike, M. M., 170
Plants for processing. *See* Processing plants
Platter, W. J., 174
Pollack, J., 166
Pordesimo, L. O., 65
Pork
  as beef substitute, 192
  fat and cholesterol levels, 194
  product recalls, 193-194
Postharvest factors, impact on sensory attributes, 167-176
  aging, 171-172
  blade tenderization, 172
  cold-shortening, 169-170, 171
  electrical stimulation, 168-170
  exsanguination, 168
  flavor, impact of aging, 172
  generally, 167-168
  immobilization, 168
  meat enhancement, 173
  stress, 168
  stunning, 168
Potassium lactate, 69, 174
Poultry
  as beef substitute, 192
  chicken contamination, 8
  fat and cholesterol levels, 194
  product recalls, 193-194
Practical Euthanasia of Cattle, 119
Prasai, R. K., 64
Preharvest beef quality
  animal identification, 122
  care and handling of beef cattle, 113-120
  extra-label drug usage, 104-105, 106-107
  foodborne pathogens, 122
  future issues, 122-123
  history of preharvest BQA, 101-103
  injection-site lesions, 102, 109-113
  over-the-counter drugs, 105

Preharvest beef quality *(continued)*
  Preharvest Beef Production Program, 101
  prescription drugs, 106
  relationship of disease to quality, 120-122
  residue avoidance, 107-109
  residue regulations, 103-105
  sensory attributes, 160-167, 161t, 162t, 163f
Preharvest control of pathogens
  direct-fed microbials, 27-30
  Escherichia coli O157:H7, 20-21
  immunomodulation, 30-33
  overview, 19-22, 33
  pathogens of interest, 22-27, 22f, 23f, 27f
Prescription drugs, 106
Price
  relative meat prices and consumer expenditures, 191-192
  signals of price system, 4
Primary processing, 40-42
Prion, 52-53
Probiotics, 27
Process attributes, 198-199
Process Verified Program (PVP), 98, 201
Processing plants
  chemical hazards, 44
  equipment hazards, 46
  further processing, 67-72
  HACCP implementation, 40
  object detection equipment, 46-47
  primary processing, 40-42
  processing interventions, 68-71
  sizes, 39-40
Producers
  changing business, 16
  in consumer product business, 16
  relationship business, 16-17
Production driven industry, 4
Production method information for consumers, 206-208
Profitability of beef producers, 3
Pseudomonas fluorescens, 57
Purcell, W. D., 189
PVP (Process Verified Program), 98, 201
Pyramid, consumer food demand, 195f

QDA. *See* Quantitative Descriptive Analysis (QDA)
QSA (Quality System Assessment), 96-97
Quality
  carcass quality. *See* Carcasses
  choice of philosophy, 87
  comparative audit results, 86, 86t
  consumer perceptions, 196-203, 202f
  cues. *See* Cues
  defined, 197
  disease and, 120-122
  documentation of, 6
  domestic market challenges, 94
  grades, 139-140. *See also* Grading of carcasses
  industry drivers, 94-95
  international standards, 91-93, 93t, 95-97
  management impact, 3
  measurement, 88
  overview, 85, 99
  preharvest. *See* Preharvest beef quality
  product attributes, 198
  product characteristics, 197-198
  quality revolution, 87-89
  Quality System Assessment, 96-97
  sensory attributes. *See* Sensory attributes
  supply chains, 98
  TQM. *See* Total quality management (TQM)
  value and, 191
Quality System Assessment (QSA), 96-97
Quantitative Descriptive Analysis (QDA)
  flavor assessment, 159-160
  flavor and texture attributes, 150-151, 152, 154
Quantity demanded, 189
Quaternary ammonium, 71

Ramsey, C. B., 206
Randel, R. D., 166
Rankings and ratings, 203-204
Ransom, J. R., 21, 60

Ready-to-eat products
cross-contamination, 47
Listeria monocytogenes contamination, 50, 70-71
processing interventions, 70
Reagan, J. O.
antimicrobial intervention, 54, 59, 60-61, 63
flavor and consumer acceptability, 160
quality grades, 140
sensory attributes, 158
tenderness attribute, 206
Recalls of product, 193-194
Record-keeping. *See* Documentation
Regulations, drug, 103-105. *See also* Legislation
Reiter's syndrome, 49
Research
Chicago beef study, 205
direct-fed microbials, 27-30
Kansas City beef study, 138
National Consumer Retail Beef Study, 128, 138-139
neomycin sulfate, 32
organic acid effectiveness, 57
ozonated water effectiveness, 59-60
Philadelphia beef study, 138, 139
quality-assurance surveys, 5
rankings and ratings, 203-204
San Francisco beef study, 138, 139, 205
sodium chlorate, 31
study of *Escherichia coli* O157:H7, 20-21, 25-26
targeted projects, 4-5
Texas A&M Ranch to Rail studies, 121-122
Restraining animals for injections, 112-113, 117
Retail operations, 71-72
Rigor mortis, 169-170
Ringkob, T. P., 170, 171
Rinsing with antimicrobial compounds, 55, 65
Ritchie, Harlan, 4
Rivera-Betancourt, M., 63
Rockwell, L. C., 165
Roosen, J., 207
Rossman, M., 60-61
Rouquette, F. M. Jr., 166
Russell, R. L., 170, 171
Safety
CAST suggestions for, 15
consumer demands, 3-6
in the food chain, 5-6
hazards. *See* Safety hazards
impact on demand, 192-194
outbreaks, 193
product recalls, 193-194
Safety hazards
biological hazards. *See* Biological hazards
chemical hazards. *See* Chemical hazards
defined, 43
HACCP. *See* Hazard Analysis and Critical Control Point (HACCP)
physical hazards. *See* Physical hazards
Salmonella
antimicrobial intervention, 58, 60, 66
control of, 8
drug resistance, 23, 24
frequency of infection, 8
generally, 6
HACCP implementation, 40
incidence and death rate, 9t, 10
Newport MDR-AmpC, 24
prevalence before and after HACCP implementation, 15, 16t
prevalence on carcasses, 10, 10t
processing interventions, 69
serotypes, 48-49
sodium chlorate control, 31
sources of, 8, 47, 48
steam pasteurization, 62
susceptible individuals, 48
symptoms, 8, 48
Salmonella Typhimurium
antimicrobial intervention, 58, 60
drug resistance, 24
immunomodulation, 31
steam pasteurization, 62
Salt, injection solution, 173
San Francisco beef study, 138, 139, 205
Sanderson, M. W., 13
Sanitation Standard Operating Procedures (SSOPs), 40, 44
Sarcomere, 169
Sargeant, J. M., 13

Savell, J. W.
antimicrobial intervention, 58
flavor and consumer acceptability, 160
grading standards, 126, 130, 132
marbling and consumer acceptance, 138, 139
organic acid effectiveness, 57
quality grades, 140
sensory attributes, 158
tenderness attribute, 170
Savell, Jeff W., 125
Scanga, J. A., 60
Schmidt, G. R., 174
Schnell, T. D., 63
Schroeder, C., 3
Schroeder, Chuck, 15-17
Schroeder, T. C., 191, 192, 194, 195, 206
Search attributes, 198
Seasonal variation in study of *Escherichia coli* O157:H7, 26
Sensory attributes. *See also* Attributes of beef
background of guidelines, 147-148
consumer evaluations, 154
defined, 147
environmental impact, 165-166
flavor. *See* Flavor attribute
high-concentrate diets, 164-165
juiciness. *See* Juiciness attribute
Meat Descriptive Sensory Evaluation, 148-150
muscle palatability, 176-179, 177t, 178t
objectives, 148
overview, 147-148, 179
panelists, 148, 151, 152, 159
postharvest factors, 167-176
preharvest factors, 160-167, 161t, 162t, 163f
Quantitative Descriptive Analysis, 150-151, 152, 154, 159-160
Spectrum Method, 150-152. *See also* Spectrum Method
temperament of cattle, 166-167
tenderness. *See* Tenderness attribute
warmed-over flavors, 150t
Sensory and Cookery Guidelines for Meat, 148
Shackelford, S. D.
antimicrobial intervention, 60-61, 63
quality grades, 142
tenderness attribute, 170, 176, 206
Shigatoxin, 48
Shigella
incidence and death rate, 9t, 10
sodium chlorate control, 31
sources of, 47, 51-52
symptoms, 51
Sickness. *See* Diseases
Silberberg, M., 143
Silberberg, R., 143
Simard, R. E., 56, 57
Siragusa, G. R., 66, 69
Skeletal maturity, 133, 140, 141-143
Slaughter
cooling the carcass, 41
evisceration, 41
exsanguination, 40-41
grading the carcass, 42, 121
hide removal, 41
inspection, 41
interventions, 41-42
physical hazards, 46
water washing, 41, 55
Slay, L. J., 166
Smith, D., 25
Smith, G. C.
antimicrobial intervention, 54, 59
beef carcass quality, 138, 139
customer vs. consumer, 4
flavor attributes, 174
grading standards, 132
injection-site lesions, 109
interventions, 60, 63, 65
Smith, R. A., 13
Smith, Robert A., 101
Smith, T. P. L., 163
Smulders, F. J. M., 170
Sodium acetate, 69
Sodium chlorate, 31
Sodium chloride, injection solution, 173-174
Sodium diacetate, 69, 175
Sodium lactate, 69, 174-175
Sodium metasilicate, 67
Sodium phosphate, 173-174
Sodium tripolyphosphate, 173
Sofos, J. N., 54, 59, 60, 63, 65
Solas, M. T., 69

South African classification system, 143
Spectrum Method
  flavor assessment, 159
  flavor and texture evaluation, 150-152
  juiciness measurement, 155
  sensory ballot, 153
Speer, N. C., 174
Spices, 46, 68
Spiderweb, 152, 154f
SSOPs (Sanitation Standard Operating Procedures), 40, 44
Staphlococcus aureus
  characteristics, 50-51
  control of, 51
  prevalence on carcasses, 10t
  sources of, 47, 50-51
Statistics, foodborne disease, 47
Steam pasteurization, 20, 62
Steam vacuuming, 61-62
Stiffler, D. M., 138
Stocker calves, 114, 116
Stopforth, J. D., 60
Storage of beef, 172
Strain Theory, 130
Stress
  heat-stress, 119-120, 166
  during immobilization, 168
  impact on quality, 166-167
Study. *See* Research
Stunning, 168
Subcutaneous administration of drugs, 110-111, 112f
Sunding, D. L., 208
Supply chains, 98
Swartz, D. R., 170, 171

Tatum, J. D.
  carcass quality and consumer acceptability, 140
  flavor attributes, 174
  flavor and consumer acceptability, 160
  injection-site lesions, 109
  pH factor, 170
  tenderness attribute, 158
Taylor, J. F., 140, 158, 160
Teams to solve quality problems, 88
Tedeschi, L. O., 166
Teeth. *See* Dentition
Temperament of cattle, impact on quality, 166-167
Temperature
  cold tolerance, 166
  cold-shortening, 164, 169-170, 171
  heat-stress, 119-120, 166
  impact on pH, 171
Tenderization, 172
Tenderness attribute
  calcium chloride ingredient, 175-176
  consumer preferences, 206
  electrical stimulation and, 169
  gene markers, 163-164
  high-concentrate diets and, 164
  overall beef tenderness, 158
  overview, 157-158
  pH factor, 170-171
Tenting the skin, 111, 112f
Tests
  Central Location Consumer Tests, 154
  In-Home Consumer Tests, 154
  Meat Descriptive Sensory Evaluation, 148-150
  National Beef Quality Audit, 94
  Quality System Assessment, 96-97
Texas A&M Ranch to Rail studies, 121-122
Texas Beef Quality Assurance Manual, 3, 14
Texture
  Flavor and Texture Descriptive Attribute Evaluation, 179
  grading, 132-133
TFQM (Total Food Quality Model), 201-202, 202f
Theories, 130
Thilmany, D. D., 208
Total Food Quality Model (TFQM), 201-202, 202f
Total quality management (TQM)
  chain reaction, 89
  overview, 89-91
  principles, 90-91
Traceability information for consumers, 206-208
Tragon Quantitative Descriptive Analysis Method, 152, 154f

Traits as important by consumers, 4-5, 5t
Transmissible spongiform encephalopathy (TSE), 52
Trimble, J., 28
Trinick, J., 173
Tronstad, R., 201
Type I muscle fibers, 176
Type IIB muscle fibers, 176-177
Typhimurium DT104, 31

Ultraviolet light, 66-67, 71
Umberger, W., 201
Umberger, W. J., 204, 207, 208
United States Agricultural Products Inspection and Grading Act, 126
United States Department of Agriculture (USDA)
  Agricultural Marketing Service, 98, 127, 200
  Food and Drug Administration, 104, 200
  FSIS. *See* Food Safety and Inspection Service (FSIS)
  history, 126
  meat-grading service, 127
  Process Verified Program, 98, 201

Vaccines
  for disease prevention, 115
  immunomodulation, 30-33
  tetanus, 115
Vacuum, steam, 61-62
Value
  creation of, 3
  quality and, 191
  value-added marketing, 97-98
  willingness-to-pay values, 207
Van Netten, P., 56
Vann, R. C., 166
VanOverbeke, Deborah L., 85
VCPR (veterinarian-client-patient relationship), 106, 109
Veal, palatability of, 167
Veld, J.H.J. Huis in 't, 56
Verocytotoxins, 48
Verotoxins, 48
Veterinarians
  extra-label drug usage, 104-105
  prescription drugs, 106
  veterinarian-client-patient relationship, 106, 109
Vibrio, incidence and death rate, 9t
Viruses
  hepatitis A, 47, 52
  Noroviruses, 47, 52
  sources of, 52
Vocalization by cattle, 117
Voges, Kristin L., 125
Vote, D. J., 174

Walton, M., 89-91
Ward, R., 201
Ware, Douglas, 29
Warmed-over flavors, 150t
Warner-Bratzler shear force, 160-161, 161t, 162t, 163, 163f
Washing of carcass during slaughter, 41, 42
Water
  cattle requirements, 114
  chlorination, 26
  during heat stress, 119
  injection solution, 173-174
  ozonated water, 59-60
  study of *Escherichia coli* O157:H7, 26
Water washing
  antimicrobial intervention, 54-55
  pressures, 65
  procedures, 54
  during processing, 41, 42
  temperatures, 65-66
Weaning, 116
Web sites
  BIFSCo, 13
  Centers for Disease Control and Prevention, 6
  HACCP principles, 14
  sources of *E. coli* O157:H7, 7
  Texas Beef Quality Assurance Manual, 3, 14
Webster, J., 52-53

Wedderburn, R. W. M., 143
Welsh, T. H. Jr., 166
Whatley, J. D., 143
Wheeler, T. L.
  antimicrobial intervention, 60-61, 63
  quality grades, 142
  tenderness attribute, 163, 176, 206
White muscle fiber, 176-177
Wilkerson, E. G., 65
Willingness-to-pay values, 207
Window of Acceptability, 130, 131f
Wise, J. W., 138, 140, 158, 160
Withdrawal time for drugs, 108-109
Womac, A. R., 65
Workforce changes, impact on demand, 195

Xiao, H., 194

Yersinia enterocolitica
  incidence and death rate, 9t
  incidence decrease, 10
  sources of, 9
  symptoms, 9
Yoder, Sally L. Flowers, 39
Younts-Dahl, S. M., 28, 29

Z lines, 169
Zepeda, C. M. Garcia, 64
Zhao, T., 27
Ziehl, A. R., 208

For Product Safety Concerns and Information please contact our EU representative GPSR@taylorandfrancis.com Taylor & Francis Verlag GmbH, Kaufingerstraße 24, 80331 München, Germany

T - #0036 - 230425 - C0 - 212/152/14 [16] - CB - 9781560223238 - Gloss Lamination